AERONAUTICAL AND ASTRONAUTICAL EVENTS OF 1961

REPORT

OF THE

NATIONAL AERONAUTICS AND SPACE ADMINISTRATION

TO THE

COMMITTEE ON SCIENCE AND ASTRONAUTICS U.S. HOUSE OF REPRESENTATIVES

EIGHTY-SEVENTH CONGRESS

SECOND SESSION

COMMITTEE ON SCIENCE AND ASTRONAUTICS

GEORGE P. MILLER, California, *Chairman*

OLIN E. TEAGUE, Texas	JOSEPH W. MARTIN, Jr., Massachusetts
VICTOR L. ANFUSO, New York	JAMES G. FULTON, Pennsylvania
JOSEPH E. KARTH, Minnesota	J. EDGAR CHENOWETH, Colorado
KEN HECHLER, West Virginia	WILLIAM K. VAN PELT, Wisconsin
EMILIO Q. DADDARIO, Connecticut	PERKINS BASS, New Hampshire
WALTER H. MOELLER, Ohio	R. WALTER RIEHLMAN, New York
DAVID S. KING, Utah	JESSICA McC. WEIS, New York
J. EDWARD ROUSH, Indiana	CHARLES A. MOSHER, Ohio
THOMAS G. MORRIS, New Mexico	RICHARD L. ROUDEBUSH, Indiana
BOB CASEY, Texas	ALPHONZO BELL, California
WILLIAM J. RANDALL, Missouri	THOMAS M. PELLY, Washington
JOHN W. DAVIS, Georgia	
WILLIAM F. RYAN, New York	
JAMES C. CORMAN, California	
THOMAS N. DOWNING, Virginia	
JOE D. WAGGONNER, Jr., Louisiana	
CORINNE B. RILEY, South Carolina	

CHARLES F. DUCANDER, *Executive Director and Chief Counsel*
SPENCER M. BERESFORD, *Special Counsel*
PHILIP B. YEAGER, *Special Consultant*
JOHN A. CARSTARPHEN, Jr., *Chief Clerk*
FRANK R. HAMMILL, Jr., *Counsel*
EARL G. PEACOCK, *Technical Consultant*
RICHARD P. HINES, *Staff Consultant*
RAYMOND WILCOVE, *Staff Consultant*
JOSEPH M. FELTON, *Publications Clerk*

FOREWORD

A chronicle of scientific and technological events in the exploration of space offers useful perspective. To those of us engaged in these activities, it provides an inventory of the crowded kaleidoscope of swift-moving domestic and foreign events. To others interested in space exploration, it helps provide a sense of pace and a clearer awareness of genuine achievements as well as greater things to come.

Events of 1961 are mingled with the past and the future. The groundwork of this year's milestones was laid several years ago. The scientific discoveries of Explorers IX, X, and XII; the suborbital Mercury flights of Alan B. Shepard and Virgil I. Grissom; the attainment of near-design speed (mach 6) and altitude (50 miles) of the X-15 rocket research airplane; the impact of Tiros satellites on global weather forecasting; and the successful first flight of the Saturn booster for large space payloads of the future—these were among the highlights of 1961. The decisions and programs undertaken this year will come to fruition in the months and years ahead.

The national character of the space program is evidenced in the contributions by American industry, the scientific community, the military services, and other Government agencies. Growing public recognition of the value of increased scientific knowledge and the ultimate benefits for society of the total space effort was also evident. Known and unpredictable promises of tomorrow spur everyone ahead in attaining the high goals of the national space program.

As our broad-based scientific program and the development of a space transportation technology underwrote the events of 1961, it was also the year in which man himself first flew in space. Such was a thrilling reminder that the best interests of all mankind must ever provide the purpose and application of space exploration.

JAMES E. WEBB,
Administrator, National Aeronautics and Space Administration.

PREFACE

This chronology represents but a first step in the historical process of fully recording and explaining the myriad activities, accomplishments, and problems of the National Aeronautics and Space Administration in the exploration and exploitation of space for the benefit of all mankind. It was prepared from open public sources. Since science and technology are fundamentally indivisible, events of space-related efforts by other governmental agencies including the Department of Defense, as well as international items of a non-NASA character, have been included to help provide the fuller context of current history. We are appreciative of the generous help of NASA offices and centers and interested members of the historical community.

This chronicle for 1961 is supplemental to "Aeronautics and Astronautics, 1915–1960," published by NASA (Government Printing Office, Superintendent of Documents). Appendix A: "Satellites, Space Probes, and Manned Space Flight, 1961," prepared by Dr. Frank W. Anderson of the NASA Historical Office, is an updating of Appendix A of "Aeronautics and Astronautics."

A chronology is but a preliminary tool of the historical process and cannot be regarded as being definitive. Historical bookkeeping and auditing are a continuous process prefatorial to full-fledged analysis. Additional comments and criticism are welcome at any time.

EUGENE M. EMME,
The NASA Historian.

CONTENTS

	Page
Foreword	III
Preface	V
January	1
February	5
March	9
April	14
May	19
June	24
July	31
August	37
September	44
October	52
November	61
December	70
Appendix A: "Satellites, Space Probes, and Manned Space Flight, 1961," compiled by Dr. Frank W. Anderson	79
Appendix B: "X-15 Flight Log," NASA release 62-95	91
Index	93

AERONAUTICAL AND ASTRONAUTICAL EVENTS OF 1961

JANUARY 1961

January 1: White House statement of President Eisenhower issued, stating that "the early establishment of a communication satellite system which can be used on a commercial basis is a national objective."

—— Project Ice Way was established near Thule by the Geophysics Research Directorate of the Cambridge Research Laboratories to test the feasibility of landing heavy aircraft on ice runways. The tests, completed in June 1961, demonstrated the strength and other engineering qualities of the ice runways constructed of natural sea water or reinforced with strands of Fiberglas.

Early January: Because of the danger of a power drain in connection with remote (tape recorder) operation, the wide-angle camera of Tiros II was turned on only for direct readout, while satellite passed over Fort Monmouth and Point Arguello.

January 3: NASA's Space Task Group, charged with carrying out Project Mercury and other manned space flight programs, officially became a separate NASA field element.

—— NASA awarded contract to General Electric for an investigation of means of storing solar heat energy in satellites.

January 4: Ablation model test with electric arc attained 4,000° F. for 105 seconds at Langley Research Center, one of a series of tests begun in September 1960.

January 5: Turbofan-powered B–52H Boeing bomber, with two prototype Douglas Skybolt air-launched 1,000-mile-range ballistic missiles under each wing, was rolled out of the factory at Wichita, Kans.

January 7: USAF Blue Scout I reached near 1,000-mile altitude with 90-pound data capsule from Atlantic Missile Range.

January 9: Jet Propulsion Laboratory awarded contract to Beckman Instruments for design studies on equipment to analyze the surface of the Moon.

—— Japanese scientist associated with Radio Research Laboratories of the Japanese Ministry of Communications began studies of space communications at NASA's Goldstone, Calif., Deep Space Tracking Station.

January 10: President-elect Kennedy received report of special nine-man committee on the national space program. Chairman of the committee was Dr. Jerome B. Wiesner of MIT.

—— A Polaris missile of the advanced A–2 design was fired from Cape Canaveral 1,600 miles down the Atlantic Missile Range. It was the third success in as many firings for the new Polaris designed to operate at a range over 1,700 miles.

January 11: President-elect Kennedy announced that Jerome B. Wiesner of MIT would be special assistant to the President for science and technology.

January 12: President Eisenhower in his state of the Union address to Congress reviewed U.S. progress in space exploration, stating, "These achievements unquestionably make us pre-eminent in space exploration for the betterment of mankind."

—— Joint DOD–NASA release outlined actions of the Aeronautics and Astronautics Coordinating Board (AACB) since its creation in September 1960.

—— First Italian launching of scientific sounding rocket in cooperative program with United States, a Nike-Cajun launched from a range in Sardinia to a height of over 100 miles, and released a cloud of sodium vapor visible for many miles.

January 13: Convair B-58 Hustler, jet bomber powered by four GE J-79 engines, broke six world speed records, Maj. H. J. Deutschendorf, U.S. Air Force, as pilot. On first closed-course run, the Hustler averaged 1,200.194 miles per hour, and it averaged 1,061.808 miles per hour on both runs carrying a payload of 4,408 pounds and a crew of three.

—— NASA announced that a Life Sciences Research Laboratory would be established on February 1 at NASA Ames Research Center, Moffett Field, Calif.

January 15: NASA began negotiations with French Commission for Spatial and Scientific Research for conducting a cooperative Franco-American space program.

January 16: FCC first allocated radio frequencies to private industry (ITT) for experiments in bouncing signals off the Moon and artificial satellites.

—— In the message of President Eisenhower accompanying his budget for fiscal year 1962, it was said: "In the program of manned space flight, the reliability of complex booster capsule escape and life support components of the Mercury system is now being tested to assure a safe manned ballistic flight into space, and hopefully a manned orbital flight in calendar year 1961. Further test and experimentation will be necessary to establish if there are any valid scientific reasons for extending manned space flight beyond the Mercury program."

—— Final assembly of first Saturn flight vehicle (SA-1) was completed.

January 17: First invention award under the authority of the National Aeronautics and Space Act of 1958 given to Dr. Frank T. McClure of the Applied Physics Laboratory of Johns Hopkins for his satellite Doppler navigation system, the $3,000 award being presented by NASA Administrator Glennan at NASA headquarters.

January 19: Report of the Space Science Board of the National Academy of Sciences stated that life in some form on other planets of the solar system may possibly exist, but that evidence of this is not available today.

—— Iris rocket, new solid-propellent single-stage sounding rocket, failed to attain programed flight from Wallops Island, reaching only 86 miles' altitude instead of 160 miles.

January 19: NASA selected Hughes Aircraft Co. for placing of a major subcontract by Jet Propulsion Laboratory to build seven Surveyor spacecraft designed for soft landings on the Moon.
—— Marshall Space Flight Center awarded contract to Douglas and Chance Vought to study launching manned exploratory expedition into lunar and interplanetary space from Earth orbits.
—— Federal Communications Commission allocated a radio frequency to the American Telephone & Telegraph Co. to establish the first space satellite communications link between Europe and the United States on an experimental basis, a program calling for NASA launching of a series of experimental communication satellites capable of relaying telephone calls, television programs, and other messages across the Atlantic.
—— NASA announced indefinite suspension of the programing of the wide-angle camera in Tiros II, the experimental weather observation satellite launched on November 23, 1960.
January 20: United States and United Kingdom signed formal agreement covering minitrack station at Winkfield, England.
—— Under NASA contract, United Technology Corp. successfully completed ground tests of three 15,000-pound thrust segmented solid-propellent rockets. Each was made up of three 1,000-pound sections which were joined prior to firing.
—— NASA's Marshall Space Flight Center (MSFC) awarded contracts to North American Aviation and Ryan Aeronautical to develop paraglider recovery system for the Saturn booster, based upon concept developed by Francis M. Rogallo of NASA's Langley Research Center.
—— Headline news in Moscow was detailed Tass announcement that Strelka, one of two female dogs recovered from orbiting Spacecraft II in August 1960, had given birth to six puppies in good health. Pravda had announced 3 weeks earlier that one of the satellite-passenger dogs had given birth.
January 23: Final test flight of USAF Atlas D traveled 5,000 miles to target down Atlantic Missile Range, representing 35 successes, 8 partials, and 6 failures in 49 test launchings for D model.
—— NASA selected United Aircraft to make feasibility study of ion rocket application for long space flights.
January 24: NASA outlined specifications for a low-altitude active communications satellite Project Relay at Goddard Space Flight Center.
January 25: NASA awarded contract to Lockheed for a spaceship refueling study.
—— NASA distributed to the world scientific community, through COSPAR, a detailed description of the next planned Beacon satellite experiment.
—— NASA revealed it had selected 12 women airplane pilots to undergo tests to determine space flight research capability.
—— Assembly of Ranger I was completed at Jet Propulsion Laboratory.
—— Titan II selected as launch vehicle for Dyna-Soar I by USAF.
January 29: NASA announced establishment of Goddard Institute for Space Studies (GISS) in New York City, which would be an extension of the Theoretical Division of Goddard Space Flight Center, Greenbelt, Md. It will be headed by Dr. Robert Jastrow.

January 30: President Kennedy stated in his state of the Union address to Congress: "This administration intends to explore promptly all possible areas of cooperation with the Soviet Union and other nations 'to invoke the wonders of science instead of its terrors.' Specifically, I now invite all nations—including the Soviet Union—to join with us in developing a weather prediction program, in a new communication satellite program, and in preparation for probing the distant planets of Mars and Venus, probes which may some day unlock the deepest secrets of the universe."

—— James E. Webb nominated as Administrator of NASA by President Kennedy.

January 30–February 2: Conference of 12 European nations held at Strasbourg to discuss a British and French proposal for a European satellite launcher development program.

January 31: USAF Samos II, a 4,100-pound test satellite containing photographic equipment, placed in orbit by Atlas-Agena A from Point Arguello, Calif.

—— Mercury-Redstone (MR–2) flight from Atlantic Missile Range shot Mercury capsule containing chimpanzee named Ham to 157 miles altitude and 418 miles down range. Capsule with life-support equipment functioned well but flight was 42 miles higher and 125 miles farther than programed. Ham was recovered in good health.

—— An eight-engine static test firing of the Saturn test booster (SA–T1) for 113 seconds was completed at Marshall Space Flight Center.

During January: International Committee on Geophysics, successor organization to the IGY, meeting in Paris, endorsed proposal for Quiet Sun Year during 1964–65. (IGY had been selected for its intense sunspot activity.)

—— NASA internal studies of a manned lunar landing program were completed. Studies considered both the direct ascent based on a large Nova-type launch vehicle and the rendezvous method of earth orbit using a number of Saturn C–2's.

—— Experiments with Echo I were discontinued except for occasional checks, having provided for innumerous communications since launch on August 12, 1960.

—— Wind tunnel testing of model of the first Saturn (SA–1) began at Arnold Engineering Development Center at Tullahoma, Tenn.

—— Explosions of Centaur engines at Pratt & Whitney led to suspension of testing.

FEBRUARY 1961

February 1: Life Sciences Laboratory established by NASA at Ames Research Center to augment, lead, direct, encourage, and coordinate biomedical research related to the space program.

—— X–15 (No. 1) flown to 49,780 feet by John B. McKay, NASA test pilot, at Edwards, Calif.

—— USAF Minuteman successful on first test launch from AFMTC, a three-stage solid-propellent ICBM with full guidance, all tested on its first launching.

—— The space surveillance system (Spasur) was formally commissioned at the Naval Weapons Laboratory, Dahlgren, Va., under the operational control of the North American Defense Command.

February 2: NASA–AEC Space Nuclear Propulsion Office invited industry to submit proposals for participation in development of Nerva (nuclear engine for rocket vehicle application), a part of Project Rover initiated in 1955 by USAF–AEC.

—— Nomination of James E. Webb to be Administrator of NASA reported favorably by the Senate Committee on Aeronautical and Space Sciences.

—— Dr. T. Keith Glennan was named consultant to the Senate Committee on Aeronautical and Space Sciences.

—— NASA announced that it would negotiate with Boeing Co., Chance Vought Corp., and Martin Co., for tanks for five first-stage Saturn launch vehicles. It later announced additional selection of Chrysler Corp.

February 4: Sputnik IV launched into orbit by U.S.S.R., a 7.1-ton payload, but mission of flight was not announced.

—— Plans to launch a Japanese Kappa 6 sounding rocket within a year announced by Yugoslavia.

February 5: Orientation of Tiros II made it impossible to obtain Northern Hemisphere pictures and malfunctions made remote picture taking undesirable, so that use of satellite's cameras was suspended until orbit precession again made Northern Hemisphere pictures possible.

February 6: NASA Aerobee-Hi successfully reached 96 miles above Wallops Station in test of behavior of liquid hydrogen in zero gravity for Lewis Research Center hydrogen propulsion development.

February 7: X–15 flown to unofficial record 2,275 miles per hour by Maj. Robert White, U.S. Air Force.

February 7–8: Meeting of NASA and contractor personnel held at NASA headquarters to review Centaur development program.

February 8: When asked at press conference about U.S. man-in-space plans, President Kennedy stated: "We are very concerned that we do not put a man in space in order to gain some prestige and have the man take a disproportionate risk * * * even if we should come in second in putting a man in space, I will still be satisfied if when we finally put a man in space his chances of survival are as high as I think they must be."

February 8: NAA delivered X-15 No. 2 with XLR-99 engine to NASA for the initiation of the NASA flight research program.

February 9: Smithsonian Astrophysical Observatory reported that Earth is a slightly irregular ellipsoid according to new calculations.

—— James E. Webb confirmed by the Senate as Administrator of NASA.

—— Gen. Thomas D. White, USAF Chief of Staff, ordered space surveillance functions transferred from Air Research and Development Command to the Air Defense Command at Ent Air Force Base, Colo., as technology in this field moved from research and development to an operational stage. The ADC established Spadats (space detection and tracking system).

February 10: Voice message sent from Washington to Woomera, Australia, by way of the Moon. NASA Deputy Administrator Dryden spoke on telephone to Goldstone, Calif., which "bounced" it to the deep space instrumentation station at Woomera. The operation was held as part of the official opening ceremony of the deep space instrumentation facility site in Australia.

—— First static test of prototype thrust chamber of F-1 engine achieved a thrust of 1,550,000 pounds for a few seconds, at Edwards, Calif.

—— Three-day meeting of Satellite Panel of the World Meteorological Organization concluded at Washington, D.C., minus participation by the Soviet member.

February 10-11: Space Science Board of the National Academy of Sciences worked out recommendation that "scientific exploration of the Moon and planets should be clearly stated as the ultimate objective of the U.S. space program for the foreseeable future." This report was submitted to the President on March 31 and was released publicly on August 6.

February 12: Sputnik VIII launched into Earth orbit by U.S.S.R., from which it placed 1,419-pound Venus probe on its course.

February 13: USAF Gam-83B, modification of Navy Bullpup, a solid-propellent air-to-surface missile, was successfully launched at supersonic speed by an F-100 Supersabre.

February 14: NASA Nike-Cajun rocket launched from Wallops Station, carrying 60-pound payload ejecting explosive charges, which fired at intervals from 20- to 80-mile altitude to provide data on density of the atmosphere.

—— Last of second series of static firings of Saturn completed at Marshall Space Flight Center for 110 seconds, approximately full duration.

—— President Kennedy congratulated Premier Khrushchev on the Soviet Union's "impressive scientific achievement" in launching a space probe toward the planet Venus.

—— NASA selected Flight Propulsion Department, General Electric, for negotiation of 18-month contract to study heatflow characteristics of fluids in nuclear powerplants.

—— NASA and United Kingdom agreed to establish joint program to test communications satellites to be launched by NASA in 1962 and 1963 in Projects Relay and Rebound.

February 15: U.S.S.R. reportedly made first photos of solar eclipse from a vehicle in space, in report later released on May 28.

—— James E. Webb was sworn in as NASA Administrator.

February 16: NASA Explorer IX placed in orbit by four-stage Scout booster from Wallops Station, the first satellite launching from Wallops, and the first satellite boosted by a solid-fuel rocket. Explorer IX was a 12-foot diameter sphere after inflation at orbital altitude.

——— NASA and France agreed to establish joint program to test communications satellites to be launched by NASA in 1962 and 1963 in Projects Relay and Rebound.

February 17: "Polka dot" Explorer IX found in orbit by visual and photographic means after failure of radio beacon delayed confirmation of orbit.

——— USAF Discoverer XX placed in polar orbit with 300-pound recovery capsule from Vandenberg Air Force Base.

——— NASA negotiated $400,000 contract with G. T. Schjeldahl Co. to design, develop, fabricate, and test rigidized inflatable spheres for Project Echo, the passive communications satellite program.

——— The last successful communication with the U.S.S.R. Venus probe was made.

February 18: USAF Discoverer XXI fired into polar orbit, and Agena B restarted in flight after first orbit.

February 20: Navy told the House Committee on Science and Astronautics that Polaris could be used as a mobile satellite launch vehicle.

February 21: NASA Space Task Group selected John H. Glenn, Jr., Virgil I. Grissom, and Alan B. Shepard, Jr., to begin special training for first manned Mercury space flight.

——— Navy Transit III-B with Lofti piggyback satellite placed into orbit by Thor-Able-Star launch vehicle, but satellites did not separate.

——— USAF canceled recovery operations of Discoverer XX capsule due to technical difficulties.

——— NASA awarded contract to G. T. Schjeldahl Co. for nine inflatable spheres for Echo program.

——— Titan ICBM completed 5,000-mile flight, the 20th success in 29 tests.

——— MA-2 launch from Cape Canaveral, trajectory providing rugged test of the Mercury capsule.

February 22: French Veronique launched capsule containing rat (Hector) to 95-mile altitude, recovered successfully.

February 23: NASA Administrator James E. Webb and Deputy Secretary of Defense Roswell Gilpatric signed letter of understanding confirming the national launch vehicle program, the integrated development and procurement of space boosters by NASA and DOD. It was agreed that neither DOD nor NASA would initiate the development of a launch vehicle or booster for use in space without written acknowledgment of the other agency.

——— Proposed DOD Directive entitled "Development of Space Systems" was submitted to the Joint Chiefs of Staff and the military services for comment by March 2.

——— Tiros II completed 3 months in orbit, continuing useful observations beyond original estimate of useful life.

February 24: NASA Juno II launched S-45 I ionosphere beacon satellite which did not achieve orbit due to malfunction shortly after booster separation.

February 25: Paul F. Bikle set world glider altitude record of 46,267 feet in Schweizer 1-23-E sailplane, beating record of 42,100 feet set by W. S. Ivans in 1950. Bikle is Director of NASA Flight Research Center, Edwards, Calif., which is conducting the X-15 flight research program.

February 26: Sputnik IV, launched on February 4, reentered the Earth's atmosphere.

February 27: FCC-NASA memorandum of understanding for delineating and coordinating civil communication space activities signed. It stated that "earliest practicable realization of a commercially operable communication satellite system is a national objective."

——— NASA released "Evaluation of U.S.S.R. versus U.S. Output in Space Science," a study prepared for the House Committee on Science and Astronautics.

February 28: NASA Administrator James E. Webb stated that President Kennedy had ordered a thorough review of the Nation's space programs.

During February: Acoustic test chamber for recording sound of rocket operations and to study human stress limits completed at Environmental Simulation Laboratory, Naval Missile Center, at Pacific Missile Range.

——— Japanese space science survey team toured NASA facilities.

——— Bell Telephone Laboratories and General Electric conducted a "phase stability" experiment on Echo I, the results indicating that the sphere was keeping its "roundness" much longer than anticipated.

——— NASA-USAF returned X-15 No. 1 to contractor (NAA) for installation of final engine of 57,000-pound thrust.

MARCH 1961

March 2: Tass announced that radio contact with the Soviet Venus probe could not be established on February 27.

—— The fourth firing of an advanced Polaris A-2, and the first from a ship, was made by the U.S.S. *Observation Island* as she cruised at 8 knots, 10 miles offshore from Cape Canaveral.

—— The President's Scientific Advisory Committee on Project Mercury visited Atlantic Missile Range for a briefing.

March 3: USAF Blue Scout II carried 172-pound payload to 1,580 miles altitude from Atlantic Missile Range.

March 6: First NASA Agena B vehicle entered checkout of systems and subsystems at Lockheed, Sunnyvale, Calif.; vehicle scheduled to launch Ranger I.

—— Department of Defense decision indicated that perfected military space vehicles would be assigned to each service which demonstrated an operational need for them, thus giving USAF major responsibility for military space development.

—— Direct-mode pictures by Tiros II camera were resumed after a month of inoperation. The quality of the pictures showed some slight improvement, supporting the theory that foreign matter may have been deposited on the lens and was gradually evaporating.

—— Equipped with turbofan engines, B-52H made its first flight at Wichita, Kans.

March 7: First flight model of Saturn booster (SA-1) installed on static test stand for preflight checkout, Marshall Space Flight Center, Huntsville.

—— NASA Marshall Space Flight Center selected Chance Vought Corp. to build 42 fuel and liquid oxygen tanks for the Saturn booster program.

—— Maj. Robert White, U.S. Air Force, flew X-15 a record speed of 2,905 miles per hour, topping his mark of 2,275 miles per hour set on February 7 with interim engine.

March 9: U.S.S.R. launched 5-ton Sputnik IV into orbit and recovered dog passenger, the second time this feat was performed.

—— Harold B. Finger was appointed Assistant Director for Nuclear Applications in NASA's Office of Launch Vehicle Programs, and continued as Manager of the AEC-NASA Space Nuclear Propulsion Office (SNPO).

—— Dr. Harold Brown, of University of California's Lawrence Radiation Laboratories, was named Director of Research and Engineering for the Department of Defense, to succeed Dr. Herbert F. York.

March 10: NASA announced first success in immediate detection in real time of radar signals off planet Venus by Jet Propulsion Laboratory Goldstone, as part of 2-month research program.

—— NASA and Navy jointly established development program to increase payload capability of Scout vehicle by 40 percent by improved performance of third- and fourth-stage engines.

—— NASA awarded contracts to Convair, Lockheed, and North American for studies of space vehicles beyond the Saturn class, having first-stage thrust of 6 to 12 million pounds.

—— National Meteorite Symposium held at Arizona State University, Tempe, Ariz.

March 13: Soviet astronomers claimed to have discovered the presence of oxygen in the atmosphere of Venus. Dr. Brian Warner of the London Observatory correlated and reinterpreted spectrographic data gathered earlier by Soviet Astronomer Nikolai Kozyrev.

—— An Atlas intended for 9,000-mile flight into the Indian Ocean plunged into the Atlantic only 200 miles from Cape Canaveral.

March 14–15: United States and United Kingdom signed formal agreement covering Mercury tracking stations on Bermuda.

Mid-March: Up to this time, approximately 78 percent of the wide-angle photographs relayed from Tiros II (weather satellite) were considered usable for current weather analysis.

March 15: NASA and United Kingdom's Space Sciences Committee agreed on experiments to be included in the second United Kingdom satellite (launched by NASA's Scout), the experiments being galactic noise, atmospheric ozone, and micrometeoroids.

March 16: Scientists from Fordham University and Esso Research announced that they had discovered waxy compounds inside a fragment of a meteorite found near Orgueil, France, in 1864.

—— NASA Robert H. Goddard Space Flight Center officially dedicated at Greenbelt, Md., dedication address delivered by Dr. Detlev Bronk, President of the National Academy of Sciences. It was the 35th anniversary of Dr. Goddard's successful launching of the world's first liquid fuel rocket. Mrs. Robert H. Goddard accepted the congressional medal honoring her husband.

March 17: Vanguard I completed third year in orbit and was still transmitting. Vanguard I provided much useful data on orbits, including the slight pear-shape of the Earth and the effect of solar pressure. Vanguard also provided the second stage for the Able, Delta, and Able-Star, as well as the third stage of Scout, pioneering solid-propellent stages used in Polaris and Minuteman.

—— First Northrop T–38 supersonic jet trainer was delivered to USAF Air Training Command at Randolph Air Force Base, Tex.

March 18: Little Joe 6 fired Mercury spacecraft from Wallops, resulted in limited test of escape system because of unprogramed sequence.

March 19: Problems with the shutter of the wide-angle camera of the Tiros II were noted, but later disappeared and did not significantly affect data from this camera.

—— Tiny particle of matter from another galaxy hit upper atmosphere of the Earth over New Mexico at a speed close to that of light and split with great force. Resultant particle shower numbered between 20 and 40 billion pieces, according to scintillation counters at the Volcano Ranch Cosmic Ray Research Center near Albuquerque, N. Mex.

March 20: Charles J. Dolan named Associate Director of NASA's Langley Research Center. He had been associated with the NASA Space Task Group since its formation at Langley in November 1958.

March 20–21: Representatives of NASA and the French Committee for Space Research agreed on cooperative space science program in meeting at Washington, D.C.

March 22: National Academy of Sciences' Geophysics Research Board announced preliminary plans for an International Year of the Quiet Sun (IQSY) during 1964–65.

—— Dr. Edward C. Welsh, a former aid to Senator Symington, was nominated by the President to be the Executive Secretary of the National Aeronautics and Space Council.

March 23: Responding to inquiry by the chairman of the House Science and Astronautics Committee, President Kennedy stated in a letter: "It is not now nor has it ever been my intention to subordinate the activities of [NASA] to those of the Department of Defense * * * there are legitimate missions in space for which the military services should assume responsibility * * * [and there are] major missions, such as the scientific unmanned and manned exploration of space and the application of space technology to the conduct of peaceful activities, which should be carried forward by the civilian space agency."

—— The first World Meteorological Day was observed by 50 nations under sponsorship of the World Meteorological Organization.

March 24: Mercury-Redstone successfully flew capsule in 115-mile flight test at Atlantic Missile Range.

—— Tiros II completed 4 months in orbit and continued to provide useful cloud picture and radiation data. Signal from Tiros II was used on 1,763d orbit to trigger dynamite to break ground for new RCA Space Environment Center at Princeton, N.J.

—— NASA and United Kingdom's Department of Science and Industrial Research signed agreement covering data acquisition unit in Falkland Islands for "topside sounder," a joint United States-Canada project.

March 25: NASA Thor-Delta fired Explorer X (P–14) into highly elliptical orbit (apogee of 148,000 miles, perigee of 100 miles) with instruments to transmit data on the nature of the magnetic fields and charged particles in this region of space where the Earth's magnetic field merges with that in interplanetary space.

—— Prof. Martin Schwarzschild, of Princeton University, named by the National Academy of Sciences to receive the Henry Draper Medal for his work as director of ONR's Project Stratoscope (produced clear photos of the structure of the surface of the Sun).

—— U.S.S.R. launched Spacecraft V, a more than 5-ton payload, and recovered capsule containing a dog named Little Star. This was apparently a repeat of the March 9 shot.

March 26: NASA Aerobee research rocket with University of Michigan payload shot to 252-mile altitude from Wallops Station.

—— Pravda article stated that the day was "not far distant when a Soviet human being will rocket into space."

March 27: Budget Director David E. Bell made known to Joint Economic Committee of Congress that the new administration would request for fiscal year 1962 $125.67 million more for NASA (in addition to previous $1,110 million) and $65 million more for the National Science Foundation (additional to $210 million).

March 27: President Kennedy initiated actions to speed up the development of large boosters.

— Dr. Carl Sagen, of the University of California, suggested that the seeding of the atmosphere of Venus with algae might alter its atmosphere to support human life.

— Its instruments recording a magnetic impulse, Explorer X became the first satellite to measure the shock wave generated by a solar flare.

March 28: USAF Dyna-Soar System Project Office personnel visited NASA headquarters for review of technical and management programs.

— President Kennedy requested Congress for $2 million so that NASA could aid FAA in development of supersonic transport aircraft. President also asked for $12 million increase in FAA budget.

— NASA Goddard scientists reported that Explorer X had encountered magnetic fields considerably stronger than expected in its elongated orbit which carried it 112,500 miles from Earth (almost halfway to the Moon), although it would take several weeks to analyze acquired data.

— Soviet press conference at Soviet Academy of Sciences in Moscow, at which Biochemist N. M. Sisakian announced that all six of Strelka's pups, on exhibit, were developing normally: "Our research on these animals, just completed, has proved that no dangerous consequences to the functioning of their organs have stemmed from the space flight. This problem has an important bearing on our preparations for man's orbiting."

— Alexander Topchiev, Vice Chairman of the Soviet Academy of Science, stated in Moscow that Western reports that some Soviet astronauts had perished in space flight attempt were "a complete fabrication * * * entirely and absolutely unfounded." Occasion was press conference at the Academy of Science on the subject of the imminent flight of man into space, at which four space dogs and six offspring were televised.

— Draft DOD directive on "Reconnaissance, Mapping, and Geodetic Programs" (5160.34), relative to development of military space systems, was sent to the services for comment.

March 29: At 280th session of disarmament conference at Geneva, Arthur H. Dean presented U.S. proposal that a system of space satellites for patrolling a ban on nuclear testing be fully operational 6 years after ratification of such a ban. Such a space patrol could "open up a new frontier of knowledge for the benefit of mankind."

March 30: Reactor-in-flight-test system (Rift) study, a part of the NASA-AEC program on nuclear rockets, was briefed by contractors at NASA headquarters.

— USAF Discoverer XXII failed to achieve orbit.

— NASA–USAF–USN rocket research X–15 flown to 169,600 feet by Joseph A. Walker, NASA pilot, the highest altitude ever reached by man and which included 2 minutes of weightlessness at the top of his climb. The X–15, powered by XLR–99 rocket engine designed to thrust it to 50 miles altitude and speeds of up to 4,000 miles per hour, was only run at three-quarters throttle.

March 30: USAF announced reduction of the B-70 program contract commitments to North American, Westinghouse, and other firms. Five major subcontracts were canceled and four others sharply reduced.

March 31: NASA selected GE's Space Sciences Laboratory and Avco Corp. for negotiation of contracts to study feasibility of magneto-gas dynamic electric rocket or thermal arc jet rocket engines.

―――― By this date, all stations of NASA's worldwide Mercury tracking network were operational.

―――― Space Science Board of the National Academy of Sciences submitted its recommendation of February 10-11 that "scientific exploration of the Moon and planets should clearly be stated as the ultimate objective of the U.S. space program for the foreseeable future."

During March: Announced that National Institutes of Health scientists were growing organisms found inside of a meteorite that fell at Murray, Ky., around 1950; first reported instance of living material, perhaps extraterrestrial, grown in a laboratory.

―――― In an experiment at Boeing, biologist J. D. McClure spent 26 hours in a sealed environment with atmosphere recycled through algae to retain 21 percent oxygen content.

―――― Marine helicopter crews conducted extensive tests to perfect water recovery of Mercury capsule at Langley Air Force Base.

―――― Personnel of NASA's technical and international programs participated in task force study of methods for increasing effectiveness of U.S. international scientific activities.

―――― It was reported that the Institute of Space Technology at Stuttgart, Germany, had developed inexpensive static test stand and fired steam motors producing 30 tons of thrust.

APRIL 1961

April 1: Dr. Charles A. Roadman named as Acting Director of the NASA Office of Life Sciences to succeed Dr. Clark Randt, who resigned effective this date.

——— Secretary of Defense McNamara issued directive (5160.34) assigning research, development, and operational responsibilities for DOD reconnaissance, mapping, and geodetic programs. The USAF was assigned responsibility for reconnaissance satellite systems as well as research and development of instrumentation and data processing associated with these satellite systems. The U.S. Army was assigned responsibility for establishment and management of a worldwide master geodetic control system, and necessary R. & D. and library support, while the USAF was made responsible for launch and recovery of geodetic payloads. The U.S. Navy was assigned responsibility for R. & D. and operation of all oceanographic and geodetic programs at sea.

——— USAF reorganized its research and development activities, creating the Air Force Systems Command (AFSC) to replace parts of the Air Research and Development Command and the Air Materiel Command, to be commanded by Lt. Gen. Bernard Schriever. Also separately created was the Office of Aerospace Research (OAR) to function as a separate air command reporting directly to Chief of Staff, USAF.

April 3: Naval Research Laboratory reported that Lofti, small piggyback satellite on Transit III-B launched on February 21, demonstrated that very low frequency radio signals pass through the ionosphere into space, thus opening new area for communications development.

April 4: Three astronauts selected for Mercury-Redstone flight (MR-3) were ordered to take refresher course in Navy centrifuge at Johnsville, Pa.

April 6: Six hundred mice placed in altitude chamber for 6-week environmental exposure at Armour Research Foundation.

——— Marshall Space Flight Center announced that 1,640,000 pounds thrust was achieved in test of F-1 rocket engine thrust chamber static firing at Edwards, Calif., a record thrust for a single chamber.

——— United States and United Kingdom signed formal agreement covering tracking station on Canton Island.

April 7-14: NASA participated in Committee on Space Research symposium held in Florence, Italy.

April 8: USAF Discoverer XXIII placed into polar orbit from Pacific Missile Range but reentry capsule stayed in orbit.

April 10: President Kennedy requested Congress to approve legislation making the Vice President Chairman of the National Aeronautics and Space Council.

April 10: Radar tracking of planet Venus for 7 weeks by Jet Propulsion Laboratory scientists had proved "astronomical yardstick" of 93,498,125 miles as the distance between the Earth and the Sun (within 1,000 miles of error).

—— Attempt to recover capsule from Discoverer XXIII unsuccessful.

—— Smithsonian Astrophysical Observatory reported that Echo I satellite may remain in orbit another 3 years.

—— Rumors swept Moscow that U.S.S.R. had placed a man into space.

April 12: U.S.S.R. announced that Maj. Yuri A. Gagarin had successfully orbited the Earth in a 108-minute flight in a 5-ton Vostok (East), the first man to make a successful orbital flight through space.

—— President Kennedy, in his regular press conference, stated that "no one is more tired than I am" in seeing the United States second to Russia in the space field. "They secured large boosters which have led to their being first in Sputnik, and led to their first putting their men in space. We are, I hope, going to be able to carry out our efforts, with due regard to the problem of the life of the men involved, this year. But we are behind * * * the news will be worse before it is better, and it will be some time before we catch up * * *."

—— USAF Blue Scout II was fired with test equipment for detection of high-altitude nuclear tests.

—— Announced in Moscow that a new State Committee for Coordinating Research Work was created, to be headed by Lt. Gen. Mikhail V. Khrunichev.

April 13: A.T. & T. stated that it hoped to orbit experimental communications satellite by May 1962, and would share use or ownership of a satellite system with other common carriers.

April 14: In response to questioning by the House Science and Astronautics Committee, Associate NASA Administrator Seamans repeated the general estimate of $20 to $40 billion as the cost for the total effort required to achieve a lunar landing, that an all-out program might cost more, and that 1967 could be considered only as a possible planning date at this stage of such a complex task.

—— Gigantic ceremony in Red Square in Moscow honoring Maj. Y. A. Gagarin, the first cosmonaut.

April 17: Subcommittee of the Senate Committee on Government Operations, Senator Hubert H. Humphrey as chairman, submitted report on "Coordination of Information on Current Research and Development Supported by the U.S. Government." It recommended innovations be developed to aid the management and conduct of research.

—— Construction of dynamic test stand for Saturn completed at Marshall Space Flight Center.

—— USAF Cambridge Research Laboratories' balloon was launched from Vernalis, Calif., maintained constant altitude of 70,000 feet for 9 days with payload of 40 pounds.

April 19: Preliminary data from Explorer X disclosed at NASA indicated that solar winds blow the Sun's magnetic field out past the orbit of the Earth.

—— Dr. Thomas Gold, of Cornell University, submitted that water exists on the Moon, perhaps shielded from evaporation by a layer of ice below the surface, in a paper given at American Physical Society.

—— Lincoln Laboratory's radar system near El Campo, Tex., began 32-day radar contact with the Sun, ending on July 7. Variations in solar activity were determined to have a corresponding effect upon the reflectivity of radio waves transmitted to the Sun from Earth.

—— Scientists from the United Kingdom and NASA announced agreement on the scientific instrumentation of a second United Kingdom satellite to be launched by NASA with a Scout vehicle.

—— Polaris fired more than 1,100 miles from submerged U.S.S. *Robert E. Lee.*

April 20: National Academy of Sciences issued report by its Space Science Board which stated that "the history of geographic exploration on Earth tells over and over again of the deaths of bold explorers. * * * To ignore this in the far more difficult and hazardous areas of man in space is foolish. Men will perish in space as they have on the high seas, in the Antarctica, in the heart of Africa, and wherever they have ventured into unknown regions."

—— House and Senate approved bill to permit Vice President of the United States to serve as Chairman of the National Space Council.

—— Dr. John R. Winckler, of the University of Minnesota, reported at the American Geophysical Union, that the first direct sampling of a cross section of solar material had revealed that particles of heavier elements of the sun ejected by solar flares have been captured in the vicinity of the Earth in study of emulsions flown by balloons and rockets during the solar activity of the fall of 1960. Tracks of helium, carbon, nitrogen, and oxygen had been detected.

April 21: USAF–USN–NASA X–15 flown to controlled-flight record speed of 3,074 miles per hour by Maj. Robert White (USAF) at Edwards, Calif. This was the first flight of X–15 with full throttle.

—— NASA fired Nike-Asp rocket carrying aluminum thermite and sodium pellets to an altitude of 34 miles, one of a series of sodium cloud firings in connection with similar launchings in Italy.

—— Dr. W. O. Roberts, Director of the National Center for Atmospheric Research (NCAR), announced that a national balloon flights facility would be established to encourage upper atmospheric research.

April 22: NASA fired first seven-stage Trailblazer rocket from Wallops Station, first three stages firing meteorite to 175-mile altitude and next four stages back through the atmosphere in a high-speed reentry experiment.

—— Italian Air Force personnel fired Jupiter IRBM in training launch at Cape Canaveral.

April 23: Tiros II completed 5 months in orbit. Useful radiation observations ceased due to detector malfunctions, but radiation electronics and tape recorder continued to function, and TV cameras continued to operate as well as on day of launch.

April 24: Dr. Leonard S. Sheingold, director of applied research at Sylvania Electronic Systems, was named by the President to be Chief Scientist, USAF.

April 25: Mercury-Atlas (MA-3) launched unmanned Mercury spacecraft in orbital test from Atlantic Missile Range, which was destroyed at 16,000 feet by range safety officer, while Mercury capsule was boosted by escape tower rockets above Atlas and subsequently recovered intact.

—— President Kennedy signed legislation making the Vice President Chairman of the National Aeronautics and Space Council.

—— Official Soviet report described preliminary weightlessness training of the Soviet cosmonauts as follows: "It was established that all selected cosmonauts possess a good ability to endure weightlessness up to 40 seconds, the cosmonaut can eat food liquid, semiliquid, and solid; can perform delicate coordinated acts, such as writing or purposeful hand motions; can maintain communication by radio; can read; and, besides, can orient himself visually."

April 27: Javelin launched 70.6-pound payload to an altitude of 475 miles in beginning of Goddard Space Flight Center program to measure the density of electrons in the ionosphere.

—— Explorer XI, a gamma-ray satellite, was successfully launched into orbit by NASA Juno II from Cape Canaveral.

—— NASA Ames Research Center measured the intensity of radiation from the hot gas over the nose of a model flying through the air at 42,300 feet per second. This speed was in excess of parabolic atmospheric entry speed and the data are significant in relation to development of lunar spacecraft. The speed, 11,100 feet per second higher than maximum air speed obtained previously, was achieved by firing the model from a light-gas gun into a highspeed jet of air flowing in the opposite direction from a shock-driven wind tunnel.

—— F. W. Reichelderfer, Chief of the U.S. Weather Bureau, testified before the House Appropriations Committee that getting the same information contained in the cloud structure photographs taken by the Tiros I weather satellite would have required thousands of weather ships over the Pacific. With Tiros I, he said, "for the first time man had a complete look at the weather over a large segment of the Earth's surface."

April 28: Little Joe 5-B launched Mercury spacecraft from Wallops Station, which provided abort test under severe atmospheric flight conditions.

—— Simulated countdown of Mercury-Redstone 3 was completed successfully.

—— First manned balloon launched from and landed back aboard a naval vessel, a Stratolab High test flight over U.S.S. *Antietam* in the Gulf of Mexico (6,000 feet).

April 28: Final NASA report on the study proposed for Saturn for use as Dyna-Soar booster was presented to the Air Force.

—— World altitude record for aircraft of 113,891 feet (34,714 meters) flown by G. Mussolov in Soviet E–66A.

April 29: Saturn booster firing of 30 seconds using timer at predetermined setting was successful in flight qualification test.

During April: The Navy reactivated the former NACA hydrodynamic research facilities at Langley Research Center, to conduct R. & D. on hydrofoils, air-cushion vehicles, hydroskis, catamarans, STOL seaplanes, torpedoes, and underwater rockets. NASA continued investigations at other facilities of Langley of ditching and water landing of space vehicles.

MAY 1961

May 1: NASA Administrator Webb issued a statement concerning the 2-year Mercury manned space flight program, which said, in part: "NASA has not attempted to encourage press coverage of the first Mercury-Redstone manned flight. It has responded to press and television requests, with the result that over 100 representatives of the press, radio, and TV are now at Cape Canaveral. * * * We must keep the perspective that each flight is but one of the many milestones we must pass. Some will completely succeed in every respect, some partially, and some will fail. From all of them will come mastery of the vast new space environment on which so much of our future depends."

—— May Day parade in Red Square, Moscow, reviewed by Maj. Yuri Gagarin beside Premier Khrushchev.

—— Tiros operations at Belmar, N.J., terminated to begin move of equipment to Wallops Station, Virginia.

May 2: Manned Mercury-Redstone (MR-3) launch postponed because of rain squalls in the recovery area.

—— USAF Bomarc B area defense missile destroyed Regulus II target missile flying at mach 2, in test at Eglin Gulf Test Range.

May 3: First silo launching of an ICBM, a USAF Titan at Vandenberg Air Force Base.

May 4: ONR Stratolab High V balloon launched from carrier *Antietam* in Gulf of Mexico reached world altitude balloon record of 113,600 feet, remaining above 104,000 feet for 2 hours 11 minutes, Comdr. Malcolm D. Ross, U.S. Naval Reserve, as pilot, and Lt. Comdr. Victor G. Prather, Medical Corps, U.S. Navy, as observer.

—— House Science and Astronautics Committee approved $126.6 million additional to the President's space budget, marking most of the increase for the Apollo program.

—— First part of MR-3 firing countdown began at T−640 minutes (7:30 a.m. eastern standard time) and held at T−390 minutes until final countdown began at 11:30 p.m. eastern standard time.

May 5: Freedom 7, manned Mercury spacecraft (No. 7) carrying Astronaut Alan B. Shepard, Jr., as pilot, was launched from Cape Canaveral by Mercury-Redstone (MR-3) launch vehicle, to an altitude of 115.696 miles and a range of 302 miles. It was the first American manned space flight. Shepard demonstrated that man can control a vehicle during weightlessness and high G stresses, and significant scientific biomedical data were acquired. He reached a speed of 5,100 miles per hour and flight lasted 14.8 minutes.

—— Saturn static firing of 44.17 seconds' duration to test-fire detection system at engine position No. 2 was successful, the second SA-1 flight qualification test at Marshall Space Flight Center.

May 5: In-house testing of Ranger I spacecraft completed at Jet Propulsion Laboratory.

May 8: Alan B. Shepard, Jr., Mercury astronaut, was awarded NASA's Distinguished Service Medal by President Kennedy in a special White House ceremony. It was followed by an informal parade to the Capitol by the seven astronauts for lunch, and a press conference at the State Department auditorium.

May 9: Senator Robert S. Kerr, chairman of the Senate Aeronautical and Space Sciences Committee, told a group at the National Radio and Television Convention that President Kennedy accepted the views of NASA and congressional leaders in approving the manned Mercury-Redstone flight of May 5.

May 9–10: Twenty-four Arcas-Robin weather sounding rockets fired within 24 hours by AFPGC at Eglin Air Force Base, Fla.

May 11: Jet Propulsion Laboratory briefed NASA headquarters on the Venus radar tracking experiment, after 2 months of intensive study begun on March 10.

—— U.S.S.R's Izvestia headlined the result of Soviet radar probes of planet Venus, a report which said that the Venusian day was from 9 to 11 Earth days, and that the astronomical unit (mean distance from the Earth to the Sun) was computed at 149,457,000 kilometers (92,812,797 miles). These figures were at variance with detailed study by scientists of JPL and MIT.

—— Static test of 111 seconds' duration of Saturn booster was successful, the final SA-1 flight qualification test of the S-I stage.

May 12: USAF announced plans to institute special course for the instruction of space pilots at Edwards Air Force Base, and it was activated in June.

May 13: NASA legislative program for the 87th Congress was submitted (S. 1857 and H.R. 7115), asking for authority to lease property, authority to acquire patent releases, elimination of the CMLC, replacement of semiannual reports to Congress with an annual one, and authority to indemnify contractors against unusually hazardous risks.

May 14: AEC's Tory II–A–1 experimental powerplant for atmospheric ramjet vehicles underwent first power tests, a part of USAF Project Pluto.

May 15: In testimony before House Appropriations Committee, Hugh L. Dryden revealed that simulated free-flight speeds just under 30,000 miles per hour had been achieved at NASA Ames Research Center, Moffett Field, Calif.

—— National Aeronautic Association announced selection of Vice Adm. William F. Raborn, Jr., to receive the Robert J. Collier trophy for his direction of the Polaris missile program.

—— Test firing of GE plug-nozzle engine developed 50,000 pounds of thrust.

May 15–17: Final reports of study contracts on Project Apollo presented by the three contractors at Langley Research Center and Space Task Group.

May 17: An HSS-2 helicopter, flown by Comdr. Patrick L. Sullivan and Lt. Beverly W. Witherspoon, set a new world class speed record of 192.9 miles per hour for 3 kilometers at Bradley Field, Windsor Locks, Conn.

May 18: First test inflation of 135-foot rigidized inflatable balloon satellite in dirigible hangar, conducted by NASA Langley Research Center and G. T. Schjeldahl Co. at Weeksville, N.C.

—— NASA selected RCA to construct the Relay experimental communications satellite to test the feasibility of transoceanic telephone, telegraph, and television communications using an active repeater satellite.

—— Announced by NASA Institute of Space Studies in New York that first major project, a 2-month seminar on the origin of the solar system, would be held in fall 1961.

May 19: Soviet Academy of Sciences revealed that the pulse rate of Maj. Yuri A. Gagarin had risen to 158 beats a minute in his Vostok flight, according to a report circulated by Tass.

—— Second Minuteman test launch was destroyed by range safety officer 90 seconds after lift.

May 19–20: Cape Canaveral opened to the general public for the first time in its history.

May 20: Unconfirmed signals were received on the frequency used by Soviet Venus probe launched February 12, according to Sir Bernard Lovell, of the Jodrell Bank Experimental Station.

May 22: Gen. Curtis E. LeMay nominated by the President to be Chief of Staff, USAF.

May 23: Tiros II completed 6 months in orbit, transmitting over 31,000 photographs of which over 75 percent have been classified as fair to good for meteorological analysis.

—— In a brief ceremony, a bust of Samuel P. Langley was presented by Paul Garber, Curator of the National Air Museum, to the NASA Langley Research Center, during which Dr. Langley's first demonstration of mechanical flight with his "Aerodrome" model in 1896 and his scientific contributions to astrophysics (i.e., the thermopile and the bolometer) were reviewed by Garber and Deputy NASA Administrator Dryden.

—— "Workshop: Telemetry in Europe" at National Telemetering Conference in Chicago brought seven European representatives together with American scientists in working out unofficial preliminary standardization planning on bands, means, and frequencies.

—— New 20-inch wind tunnel at the Aeronautical Research Laboratory at Wright-Patterson Air Force Base announced as capable of testing at mach 14, at 200,000-foot altitude, and at 2,500° F.

May 24: FCC endorsed the ultimate creation of a commercial satellite system to be owned jointly by international telephone and telegraph companies and announced that it was calling a meeting on June 5 to explore "plans and procedures looking toward early establishment of an operable commercial communication satellite system."

—— Launching of NASA ionosphere beacon satellite (S–45 II) at Atlantic Missile Range unsuccessful when Juno II power supply failed and prevented ignition of second stage.

—— Operational Atlas raised from emplacement and fired in an operational test exercise at Vandenberg Air Force Base.

May 24: Three Navy F4H Phantom II fighters, competing for the Bendix Trophy, bettered the existing record for transcontinental flight from Los Angeles to New York. The winning team of Lt. R. F. Gordon, pilot, and Lt. (jg.) B. R. Young, RIO, averaged 870 miles per hour on the 2,421.4-mile flight and established a new record with a time of 2 hours 47 minutes.

—— Comdr. P. L. Sullivan, U.S. Navy, and Lt. B. W. Witherspoon, flying an HSS-2 helicopter, set another new world class speed record with a mark of 174.9 miles per hour over a 100-kilometer course between Milford and Westbrook, Conn.

—— National Rocket Club President H. A. Timken announced proposal to Secretary of the Treasury Dillon to consider a special series of savings bonds and savings stamps to finance the U.S. space program, to be known as series S bonds for space.

May 25: In his second state of the Union message President Kennedy reported to Congress regarding the space program: "With the advice of the Vice President, who is Chairman of the National Space Council, we have examined where we [United States] are strong and where we are not, where we may succeed and where we may not. * * * Now is the time to take longer strides—time for a great new American enterprise—time for this Nation to take a clearly leading role in space achievement which in many ways may hold the key to our future on Earth." President Kennedy set forth an accelerated space program based upon the long-range national goals of landing a man on the Moon and returning him safely to Earth; early development of the Rover nuclear rocket; speed up the use of Earth satellites for worldwide communications; and provide "at the earliest possible time a satellite system for worldwide weather observation." An additional $549 million was requested for NASA over the new administration's March budget requests; $62 million was requested for DOD for starting development of a solid-propellant booster of the Nova class.

May 25: At NASA press conference following President Kennedy's call to Congress for an accelerated space program, NASA Administrator Webb pointed out that the long-range and difficult task of landing a man on the Moon and returning him safely to Earth before the end of the decade offered a chance to beat Russia.

—— X-15 flown to record speed of 3,300 miles per hour by NASA test pilot, Joseph Walker, at Edwards Air Force Base.

—— Kaman H-43-B Huskie helicopter flown to claimed altitude record of 25,814 feet by Capt. W. C. McMeen (USAF), bettering Russian record of 24,491 feet established on March 26, 1960.

—— Prerecorded voice message successfully transmitted from NRL to BTL via Echo I, the quality of the transmission being virtually as good as previous experiments.

May 26: USAF B-58 Hustler flown from Carswell Air Force Base, Tex., to Le Bourget, Paris, in record 6 hours 15 minutes, covering distance from New York to Paris in 3 hours 20 minutes. This flight commemorated the 34th anniversary of Charles A. Lindbergh's transatlantic crossing on May 20-21, 1927, and the opening of the 24th Paris International Air Show.

—— Atlas E fired successfully from Atlantic Missile Range.

May 26-27: First National Conference on the Peaceful Uses of Space, held at Tulsa, Okla., at which leading American space scientists and technologists appraised the current and future applications of space science and technology for human welfare. It was sponsored by NASA and the Tulsa Chamber of Commerce, with the Aerospace Industries Association, Aerospace Medical Association, American Astronautical Society, American Institute of Biological Sciences, the American Rocket Society, the Electronic Industries Association, Frontiers of Science Foundation (Oklahoma), and the Institute of the Aerospace Sciences as cosponsors.

May 26-June 4: Freedom 7, Mercury spacecraft in which Alan B. Shepard, Jr., made his space flight on May 5, was a major drawing card at the Paris International Air Show. Details of the spacecraft and of Shepard's flight were related to about 650,000 visitors.

May 27: Dr. Lloyd V. Berkner, Chairman of the Space Science Board of the NAS, stated: "Since, as space activity becomes more difficult and advanced, the space effort will be limited by our knowledge of space at any time, leadership in space science must soon become one of the controlling factors in acquiring space leadership generally." Berkner spoke at the first National Conference on the Peaceful Uses of Space held at Tulsa, Okla.

May 29: Atlas booster 111-D, to be used for Ranger I, was erected on the launch pad at Cape Canaveral.

May 30: U.S.S.R. revealed first details concerning Cosmonaut Gagarin's orbital space flight on April 12, when application was made to the International Aeronautical Federation (FAI) to have flight made an official world record: Duration, 108 minutes; maximum altitude, 203 miles; launch site, cosmodrome at Baikonur (near Lake Aral); landing site, near village of Smelovka in Seratov region; launch booster, six-engine rocket with total boost of 20 million horsepower.

May 31: Three-week meeting of the executive council of the U.N. World Meteorological Organization in Geneva concluded, at which 18 national representatives (including A. A. Zolotoukhin of the U.S.S.R.) discussed general basis for international use of weather satellites. Dr. Francis W. Reichelderfer, Chief of the U.S. Weather Bureau, was a U.S. representative.

During May: Army Chemical Corps Biological Laboratories completed preliminary tests of microorganisms in a simulated space vacuum at the National Research Corp.

—— Complete system studies of the Apollo spacecraft system that were begun in November 1960 were completed by three industrial contractors.

JUNE 1961

June 1: NASA awarded contract for developing means of sterilizing space vehicles to Wilmot Castle Co.

—— AEC and NASA jointly announced plans for Kiwi B reactor test at Jackass Flats test site in Nevada.

—— NASA announced that a two-stage Saturn C-1 will be used for the first 10 research and development flights.

June 2: Collapse of a lock in the Wheeler Dam below Huntsville on the Tennessee River interdicted the planned water route of the first Saturn space booster from Marshall Space Flight Center to Cape Canaveral on the barge *Palaemon*.

—— Deputy Premier Mikhail Khrunichev, chief coordinator of the Soviet Union's man-in-space program, died in Moscow.

June 3: Dr. Edward R. Sharp, former Director of Lewis Research Laboratory (1942-61), was presented NASA's first Outstanding Leadership Medal by Dr. Hugh L. Dryden.

—— USAF B-58 which established Atlantic crossing record to Paris of 3 hours 19 minutes crashed after takeoff from Le Bourget Airport, killing its three-man crew. Maj. Elmer E. Murphy, pilot, had recently been awarded the Louis Bleriot Speed Trophy for record speed flight of 1,302 miles per hour in January.

—— Aerojet-General test-fired large solid-propellent rocket motor which generated a half million pounds' thrust, at Sacramento, Calif.

—— A leading Istanbul newspaper, Milliyet, reported Turkish newsmen's reactions after seeing movies of both the Shepard and Gagarin space flights: "When the film was over the journalists asked the Soviet consul general: 'In the Shepard film we followed all phases of his space flight, but in yours we followed only Khrushchev. * * * Why don't you show your space flight too?' The Tass correspondent on behalf of the consul general answered: '* * * We are mainly interested in people's excitement and reaction. This is what we wanted you to see.' Gagarin may have gone into space, but this is not the impression of the journalists who saw both films: Shepard really went into space, not Gagarin, and in front of the whole world, too."

June 4: Nationwide Gallup poll released which showed that 38 percent of those questioned thought the United States led in space research, while 38 percent thought the U.S.S.R. led. The same balance also was tabulated on which Nation will be the first to place a man on the Moon.

—— Northrop disclosed "porous wing" plane under development for USAF, modified version of WB-66D based on inhalation concept (eliminating up to 80 percent of the frictional drag) proposed by Werner Pfenninger. Work on drag reduction by means of increasing the laminar flow by boundary layer suction had been performed at Langley Aeronautical Laboratory in the late 1930's by Albert E. Doenhoff and Ira H. Abbott.

June 5: Huge Saturn launch complex at Cape Canaveral dedicated in brief ceremony by NASA, construction of which was supervised by the Army Corps of Engineers. Giant gantry, weighing 2,800 tons and being 310 feet high, is largest movable land structure in North America.

—— Two pilots sealed in 8- by 12-foot simulated space cabin for 17-day round trip to the Moon, at the School of Aerospace Medicine, San Antonio, Tex.

June 6: Biomedical results of Mercury-Redstone space flight of Alan B. Shepard, Jr., publicly reported at a special conference in Washington sponsored by NASA, National Institutes of Health, and the National Academy of Sciences. Shepard's heart reached a maximum of 138 beats per minute during the flight.

—— NASA Agena B management meeting was held at Marshall Space Flight Center with representatives from MSFC, NASA headquarters, AFSSD, LMSD, JPL, and GSFC.

—— USAF Aerobee-Hi with Cambridge Research Laboratory payload designed to trap space dust, reached 101 miles over White Sands Missile Range.

June 7: In address at George Washington University, NASA Administrator Webb stated that the exploration of space was an important part of man's "driving, restless, insatiable search for new knowledge."

—— Research Analysis Corporation established by U.S. Army as a nonprofit advanced research organization to replace ORO of Johns Hopkins University.

—— AEC-NASA jointly announced plans to negotiate with an industrial team for a first-phase contract for the development of the Nerva nuclear rocket engine. Team selected for the Nerva part of Project Rover consisted of Aerojet-General Corp. and Westinghouse Electric Corp.

—— NASA Administrator James E. Webb announced creation of a new Office of Programs to be headed by D. D. Wyatt, and the renaming of the Office of Administration under Albert F. Siepert.

June 8: Small rocket lift device demonstrated publicly for the first time at Fort Eustis, Va., a rocket belt developed by Bell Aerosystems, which lifted Harold M. Graham in a controlled free flight to an altitude of 15 feet and a standup landing 150 feet from his starting point.

—— USAF Discoverer XXIV failed to achieve orbit.

—— NASA announced accelerated recruiting of qualified scientists and engineers at its field centers to fill anticipated manpower requirements in the expanded space exploration program. During 1960 NASA interviewed 3,000 persons on 100 college campuses.

—— Astronomers of Lick Observatory positioned 36-inch refractor telescope so as to intersect the path of Echo I at its predicted point of maximum elevation. Prediction of Goddard Space Flight Center was confirmed at exact time and within 10 minutes of arc.

June 9: NASA press conference revealed that data from Vanguard III (during November 15–17, 1960) and Explorer VIII (also during November 1960) indicated that high-velocity clouds of micrometeorites moved near the Earth, perhaps in a meteor stream around the Sun. This new data was just revealed from completed analysis.

——— Echo I completed its 3,697th orbit after 9 months. When this first passive communications "balloon satellite" was launched on August 12, 1960, it was not expected to have a long life span.

June 10: National Bureau of Standards broke ground for new research facility at Gaithersburg, Md., which will include a megapound deadweight testing machine to provide measurement standards for multimillion rocket thrust requirements.

——— NASA Ad Hoc Task Group, created on May 25 to survey launch vehicles and their development schedules pertinent to the manned lunar landing program, reported its findings.

June 12: British and Soviet scientists were still unable to identify signals received since May 17 as being transmitted from the Soviet Venus probe launched on February 12.

——— Reuters reported that the United States pays about $140 per hour for use of the Jodrell Bank Observatory in England, while the U.S.S.R. pays nothing. Sir Bernard Lovell explained that "the Americans occupy the telescope for long periods, where the Russians scarcely use it."

——— NASA's Incentive Awards Committee determined that Dr. Henry J. E. Reid, Director Emeritus of the Langley Research Center, would receive NASA's Outstanding Leadership Medal.

June 13: NASA Engineer Test Pilot Joseph A. Walker, who hit record altitude of 169,600 feet on March 30 and record speed of 3,300 miles per hour on May 25 in the X–15, received the 1961 Octave Chanute Award at IAS meeting in Los Angeles.

June 13–25: Freedom 7 Mercury capsule displayed to approximately 750,000 visitors at the Rassegna International Electronic and Nuclear Fair at Rome, Italy.

June 14: NASA's Plum Brook nuclear test reactor at Sandusky, Ohio, went critical for the first time. This reactor was begun in September 1956, and the facility presently has a staff of 100 persons, headed by Dr. Theodore M. Hallman.

——— NASA and the Argentine Comision Nacional de Investigaciones Espaciales signed a memorandum of understanding for a cooperative space science research program using sounding rockets.

——— Four-stage Javelin fired to 560-mile altitude from Wallops Island, testing extension of two 75-foot antenna arms on radio command at altitude, a test flight in the United States-Canadian Alouette satellite development.

June 15: Search for U.S.S.R. Venus probe "lost" since February was ended at Jodrell Bank radiotelescope, as visiting Soviet space scientists, Alla Masevitch and Jouli Khodarev, prepared to leave. The U.S.S.R. Venus probe was last commanded on February 12.

——— President Kennedy presented the Robert J. Collier Trophy to Vice Adm. William F. Raborn, Jr., who had directed the development to the Polaris IRBM.

June 15: President Kennedy directed the National Aeronautics and Space Council to undertake a full study of the Nation's communications satellite policy, stated that leadership in science and technology should be exercised to achieve worldwide communications through the use of satellites at the earliest practicable date. While no commitments as to an operational system should be made, the President stated that the Government would "conduct and encourage research and development to advance the state of the art and to give maximum assurance of rapid and continuing scientific and technological progress."

June 16: USAF Discoverer XXV placed into polar orbit by Thor-Agena B, at Vandenberg Air Force Base, Calif.

—— NASA Ad Hoc Task Group, established to determine the main problems, the pacing items, and the major decisions required to accomplish the manned lunar landing mission, reported its findings. The direct ascent mission was used in this intensive study with less detailed consideration of the rendezvous method.

June 18: Senate Government Operations Subcommittee on National Policy Machinery released report on "Science Organization and the President's Office." This study recommended that a new Office of Science and Technology be created in the Executive Office of the President.

—— Skindivers parachuted north of Hawaii to recover the capsule of Discoverer XXV, which carried samples of common and rare metals.

—— Presidium of the U.S.S.R. Supreme Soviet awarded 7,026 honors to those associated with the flight of the spaceship satellite Vostok I: Nikita S. Khrushchev received the Order of Lenin and a third Gold Hammer and Sickle Medal for "guiding the creation and development of the rocket industry, science, and technology" which "opened up a new era in the conquest of space"; 7 outstanding scientists and designers received a second Gold Hammer and Sickle Medal; 95 designers, officials, and technicians received the title of Hero of Socialist Labor; and 6,924 workers, designers, scientists, and technicians received various orders and medals (Order of Lenin, 478 persons; Order of the Red Banner of Labor, 1,218; Order of the Red Star, 256; Order of the Badge of Honor, 1,789; and medals to 3,183 other persons).

June 19: Harmon International Aviator's Trophy for 1961 announced as going to three winners for the first time—X-15 rocket research airplane pilots: A. Scott Crossfield, of North American; Joseph A. Walker, of NASA, and Maj. Robert A. White, U.S. Air Force.

—— NASA announced contract with the National Research Corp. to determine whether six types of microbes can sustain simulated exposure to the space environment including ultrahigh vacuum, ultraviolet radiation, and fluctuating temperatures.

—— Yuri Gagarin reported in Pravda that "I was in the center of a whirl of flames" when his Vostok spacecraft reentered the atmosphere on April 12. His book, "Road to Outer Space," was being serialized in Pravda.

—— Legislature of the State of Alabama considered investment of $3 million in establishing a Space Research Institute at Huntsville as a joint University of Alabama and Auburn University center.

June 20: Nuclear Vehicles Project Office established at Marshall Space Flight Center, Col. Scott Fellows, U.S. Air Force, named as Chief.

June 21: Five-year agreement on scientific cooperation signed in Moscow by representatives of the academies of science of the U.S.S.R. and Red China, according to Tass.

——— NASA Administrator Webb accepted one of the three President's Safety Awards for accident prevention during 1960. He pointed out that NASA's activities involved test flying of experimental aircraft, untried highly explosive fuels, high-voltage electricity, and highly pressurized air and superheated temperatures, in addition to rocket and spacecraft tests and launching and the operation of two nuclear reactors and a cyclotron.

——— Hypersonic wind tunnel at Douglas Aircraft became operational at El Segundo, reportedly the largest industry-owned tunnel in the United States (36 inches long, 6-inch diameter, capable of mach 10).

——— USAF Mace B made 1,100-mile guided flight, ending its R. & D. phase.

June 22: Deputy NASA Administrator Dryden sent an explanatory letter to Chairman Robert S. Kerr, of the Senate Committee on Aeronautical and Space Sciences, on the broad scientific and technological gains to be achieved in landing a man on the Moon and returning him to Earth. Dr. Dryden pointed out that this difficult goal "has the highly important role of accelerating the development of space science and technology, motivating the scientists and engineers who are engaged in this effort to move forward with urgency, and integrating their efforts in a way that cannot be accomplished by a disconnected series of research investigations in several fields. It is important to realize, however, that the real values and purposes are not in the mere accomplishment of man setting foot on the Moon but rather in the great cooperative national effort in the development of science and technology which is stimulated by this goal." Dr. Dryden pointed out that "the billions of dollars required in this effort are not spent on the Moon; they are spent in the factories, workshops, and laboratories of our people for salaries, for new materials, and supplies, which in turn represent income for others * * *. The national enterprise involved in the goal of manned lunar landing and return within this decade is an activity of critical impact on the future of this Nation as an industrial and military power, and as a leader of a free world."

——— Mercury-Redstone booster for MR-4 flight was erected on pad 5 at Atlantic Missile Range.

——— K. Kordylewski, of the Cracow Observatory in Poland, was reported to have photographed two cloudlike objects, possibly natural satellites of the Earth.

June 23: NASA–DOD Executive Committee for Joint Lunar Study and a Joint Lunar Study Program Office established by letter directive to work out and define support requirements for the U.S. manned lunar landing program.

June 23 NASA–USAF–USN X-15 flown to 3,603 miles per hour (mach 5.3), record for manned aircraft by Maj. Robert White, U.S. Air Force, which was faster than a mile per second. Losing cabin pressure at 100,000 feet, White was able to pilot the X-15 safely because of full-pressure suit. This was the fifth powered NASA flight with the XLR-99 engine.

—— Joint study was undertaken by NASA and DOD to make recommendations on the launch site to be used for the manned lunar exploration missions. A report of this study was completed in July.

—— Nike-Cajun sounding rocket fired from Eglin Gulf Test Range by Cambridge Research Laboratory with micrometeorite counting payload.

—— Director of Marshall Space Flight Center directed that further engineering work on Saturn C-2 configuration would be discontinued, and that efforts would be applied to clarification of the Saturn C-3 and Nova concepts.

—— Tiros II completed 7 months in orbit, still providing useful data.

June 24: President Kennedy assigned Vice President Johnson the task of unifying the Nation's communications satellite programs, in a letter which stressed urgency and "highest priority" for the public interest.

—— Mercury capsule was modified for MR-4 flight, with observation window replacing two viewports and with improved manual control system.

June 26: In an interview in U.S. News & World Report, NASA Administrator Webb stated that "the kind of overall space effort that President Kennedy has recommended * * * will put us there [on the Moon] first." This achievement, costing between $20 and $40 billion, "probably toward the $20 billion level * * * will be most valuable in other parts of our economy." Mr. Webb said that the U.S.S.R. did have an advantage in being able first to orbit a multimanned spacecraft around the Earth and also around the Moon.

—— A Navy YFNB barge was obtained by NASA to serve as a replacement for the *Palaemon* in transporting of the Saturn booster to Cape Canaveral.

June 27: Senate Aeronautical and Space Sciences Committee unanimously approved the administration's $1,782,300,000 budget for NASA in fiscal year 1962.

—— Eberhardt Rechtin, of Jet Propulsion Laboratory, questioned the Soviet calculations on the rotation speed of the planet Venus and the astronomical unit, and suggested that the Soviet scientists may have been influenced by earlier MIT studies (1958). Completion of extensive radar studies of Venus by Jet Propulsion Laboratory Goldstone, he submitted, provided more accurate information. The differing figures as reported are—

    U.S.S.R (1961): 9 to 11 days' rotation—A.U. 92,812,797 miles.

    Jet Propulsion Laboratory (1961): About 225 days' rotation—A.U. 92,956,000 miles.

—— Eight-engine static test of Saturn SA-T2 of 29.9 seconds' duration successful at Marshall Space Flight Center.

June 28: First showing of new Soviet aircraft in flight rehearsal for an air show on July 9 in Moscow (first major air show since 1956), one a large delta-wing jet bomber perhaps comparable to the B-58, as well as a turboprop Bear Tu-114 carrying missiles.

June 28–July 21: A Planning Task Force of the National Academy of Sciences Committee on the Atmospheric Sciences met in a series of six separate conferences in Boston to lay out a 10-year plan to guide long-range use of Government research funds.

June 29: First launching of three active satellites in one shot, and the first launching of a satellite with nuclear power, when a Thor-Able-Star launched Transit IV-A (equipped with an atomic radioisotope-powered battery of the Snap series), and two accompanying satellites, Injun and Greb III, from Atlantic Missile Range. Transit IV is forerunner of a navigation satellite system, while Injun gathers data on the radiation belts, and Greb III gathers data on X-ray radiation from the Sun.

—— NASA awarded contract to Pratt & Whitney for development of space radiators and condensors for the Lewis Research Center.

June 30: In Scout launching of micrometeorite counter satellite (S-55) from Wallops Station, third stage did not ignite, and the vehicle was destroyed.

—— Dr. Henry J. E. Reid, senior staff associate and former Director of the Langley Research Center, retired after over four decades of Government service. He began as a junior engineer at Langley in April 1921, became Director in 1926, in which capacity he served for 34 years.

—— Navy announced that Injun and Greb satellites placed in orbit with Transit IV-A had not separated and were thus not functioning at full efficiency.

During June: National Academy of Sciences established the Geophysics Research Board (GRB) in 1960 in response to a request from the International Council of Scientific Unions (ICSU). By June 1961, it had four active panels to consider specific international programs: World Magnetic Survey (WMS); International Year of the Quiet Sun (IQSY); International Exchange of Scientific Data; and Solid Earth Problems.

—— Dr. von Kármán and some of his associates organized the Astronautics Foundation, Inc., in Washington, D.C., to enable U.S. individuals and corporations to support through this nonprofit foundation various cooperative international activities.

—— Boeing began modification of B-52 to carry aloft and release the Dyna-Soar manned space glider.

—— Army Redstone missile completed its 8-year military test program (41 successes in 45 launchings).

—— NASA entered letter contract with RCA for four additional Tiros weather satellites to extend the program.

JULY 1961

July 1: Weather Bureau announced that cloud cover pictures taken by Tiros I went on public sale at the National Weather Records Center, Asheville, N.C.

—— The space detection and tracking systems (Spadats) began NORAD operations as scheduled, a system which "detects, tracks, and identifies manmade objects in space and consolidates and displays information regarding such objects."

—— First anniversary of Marshall Space Flight Center as a NASA center, and NASA Administrator Webb visited the Huntsville facility.

—— U.S. Air Force reorganized its headquarters staff to reflect creation of Air Force Systems Command (AFSC), which made the Deputy Chief of Staff for Development, Lt. Gen. Roscoe C. Wilson, Deputy Chief of Staff for Research and Technology.

July 5: NASA announced awarding of study contract of Douglas Aircraft Co. for the development of orbital placement techniques and engineering design for Project Rebound inflatable spheres.

—— NASA awarded contract with Boeing to investigate the development of large Saturn-Nova class rockets employing different combinations of liquid or solid types of fuel.

—— Israel fired a multistage solid-propellent Shavit (Meteor) II rocket to an altitude of 80 kilometers, releasing a sodium-cloud meteorological payload.

—— Major Gagarin, speaking in Helsinki, Finland, stated that the U.S.S.R. would launch another manned space vehicle sometime before the end of the year.

—— French Foreign Ministry announced that West German Government had agreed in principle to participate in 12-nation program to construct launch satellites.

July 7: USAF Discoverer XXVI orbited from Vandenberg Air Force Base, carrying instrument capsule to be recovered after 32 polar orbits and 4 days. Of the 25 previous Discoverers, 17 had gone into orbit and 15 carried recovery capsules, of which 5 had been recovered.

—— The second static firing of the Saturn SA–T2 test booster was successfully completed at Marshall Space Flight Center in an eight engine test of 119 seconds' duration.

—— Atlas E, launched from Cape Canaveral, established distance flight record of 9,050 miles, its nose cone landing 1,000 miles southeast of Cape Town, South Africa.

July 9: Massive Soviet air show over Tushino Airport in Moscow, on Soviet Air Force Day, which demonstrated that U.S.S.R. had continued development of all classes of military aircraft.

—— Capsule of Discoverer XXVI snatched at 15,000 feet during final descent after 32 polar orbits. Midair recovery by C-119, Capt. Jack Wilson, U.S. Air Force, as pilot, was fourth so performed. Capsule carried undisclosed payload.

July 9: National Science Foundation released forecast of the Nation's science needs for the next decade, which predicted that United States would need nearly twice as many scientists in 1970 (168,000) as today (87,000).

—— Reported that Navy had been launching telephone poles with rocket boost in test of floating launching requirements.

July 10: National Science Foundation policy document entitled "Investing in Scientific Progress" was released, which showed dollar and manpower investments needed by United States in decade 1960–70 to ensure fulfillment of the Nation's research capabilities.

July 11: NASA announced that a complete F–1 engine had begun a series of static test firings at Edwards Rocket Test Center, Calif.

July 11–12: Cosmonaut Gagarin visited England.

July 12: NASA Tiros III weather satellite successfully launched into near-circular orbit by Thor-Delta from Cape Canaveral.

—— Midas III (missile defense alarm system) launched into polar orbit from Pacific Missile Range, with record 1,850-mile-high orbit and was heaviest U.S. satellite launched to date. Second-stage Agena B was restarted at apogee of first orbit.

—— Jet Propulsion Laboratory announced that construction was underway on the first large space simulator in the United States capable of testing full-scale spacecraft of the Ranger and Mariner classes with the three primary space effects—solar radiation, cold space heat sink, and a high vacuum equivalent to about one part in a billion (1:1,000,000,000) of the atmospheric pressure on the Earth.

July 13: Mercury-Redstone 6 was static tested for 30 seconds at Marshall Space Flight Center to ensure satisfactory operation of the turbopump assembly.

July 13–14: Two Nike-Cajun rockets launched University of New Hampshire-Goddard Space Flight Center payloads from NASA Wallops Station.

July 14: Advanced Polaris fired 1,600 miles down Atlantic Missile Range with all-inertial guidance system.

—— Simulated Mercury-Redstone 4 flight test held at Atlantic Missile Range.

July 16: Vice President Johnson announced that the National Aeronautics and Space Council had reached unanimous agreement on the national communications satellite policy, and unspecified recommendation transmitted to President Kennedy.

July 17: NASA announced selection of RCA Astro-Electronics Division to build seven capsules for experimental ion propulsion engines.

—— A joint tenancy agreement for NASA and DOD use of the Atlantic Missile Range was signed by Commander, Atlantic Missile Range, and the Director of Launch Operations, NASA.

July 18: FAI (Fédération Aeronautique Internationale) officially recognized the first space flight records claimed by U.S.S.R. and the United States:

Yuri Gagarin (April 12, 1961): Duration in orbital flight, 108 minutes; greatest altitude in Earth orbital flight, 203 miles; greatest mass lifted in Earth orbital flight, 10,395 pounds.

AERONAUTICAL AND ASTRONAUTICAL EVENTS OF 1961

Alan Shepard (May 5, 1961: Altitude without orbit, 115.696 miles; greatest mass lifted without Earth orbit, 4,031.7 pounds.

—— United States-U.S.S.R. talks began on bilateral agreement on commercial air flights between New York and Moscow.

—— Saturn SA-T2 booster successfully static tested for 111 seconds at Marshall Space Flight Center.

July 18–20: NASA-Industry Apollo Technical Conference held in Washington, D.C., which assembled Apollo requirements with participation of Space Task Group, representatives of other NASA Centers, and the three Apollo study contractors—General Dynamics/Astronautics, General Electric, and Martin.

July 19: Mercury-Redstone (MR-4) with manned Liberty Bell 7 capsule canceled within minutes of launch because of adverse weather.

—— NASA and Weather Bureau invited over 100 of the world's weather services to participate in the Tiros III satellite experiment by conducting special ground-based observations synchronized with passes of the satellite.

—— Tiros III photographed tropical storm Liza in the Pacific Ocean, pinpointing its location for meteorologists.

July 20: NASA and DOD, following an exchange of letters between the Administrator of NASA and the Secretary of Defense, established a joint study to determine the national large launch vehicle needs for the next decade, considering the requirements of both NASA and DOD.

July 21: MR-4 Liberty Bell 7, manned by Mercury Astronaut Virgil I. Grissom, made successful 15-minute, 118-mile-high and 303-mile-long flight down Atlantic Missile Range, premature blowout of escape hatch flooding capsule and making helicopter pickup of Grissom difficult. Capsule sank in 18,000 feet of water after warning light indicated helicopter engine was overheating and the capsule was cast loose. This was the second successful manned suborbital space flight.

—— President Kennedy signed NASA's fiscal year 1962 authorization bill providing for a total of $1,784,300,000.

—— USAF Discoverer XXVII destroyed by range safety officer 60 seconds after launch from Vandenberg Air Force Base.

—— At the request of Senator Paul H. Douglas, the membership of the American Astronautical Society was polled by the University of Illinois Observatory as to their opinion regarding the "scientific value" of the U.S. space program to land on and return one or more men from the Moon.

—— USAF processed 44 test pilots through the Dyna-Soar selection program, and reportedly both USAF and NASA test pilots would participate in the step I suborbital flight program.

July 22: Astronaut Virgil Grissom was awarded the NASA Distinguished Service Medal by Administrator Webb at conclusion of MR-4 press conference at Cape Canaveral.

July 23: Red Star (Krasnaya Zvezda) of the U.S.S.R. stated Tiros III and Midas III launched on July 12 were comparable to the U-2: "A spy is a spy, no matter what height it flies."

July 23: NASA Administrator Webb, in congressional testimony, pointed out that the Tiros cloud-cover program was known to the entire world, involved no surveillance, and promised great benefit to all nations. He pointed out that data from Tiros satellites had been made available to all, including the Soviet Union.

—— Central Aero Club of the U.S.S.R., in seeking to place Gagarin's flight in the record books, revealed to FAI Astronautics Documentation Subcommittee, meeting in Paris, that Gagarin rode his spacecraft to Earth rather than parachuting.

July 24: White House issued statement by President Kennedy on "Communication Satellite Policy," which outlined governmental responsibilities for research and development "to give maximum assurance of rapid and continuous scientific and technological progress," and which affirmed that "private ownership and operation of the U.S. portion of the system is favored" within eight policy requirements. The President's statement said that through this country's leadership, communications through the use of space satellites should be developed "for global benefit at the earliest practicable date." He invited "all nations to participate in a communication satellite system in the interest of world peace and closer brotherhood among peoples throughout the world." And, during present phase of research and development, "no arrangements between the Government and private industry [should] contain any commitments as to an operational system." In conclusion, the President said that "I am anxious that development of this new technology to bring the farthest corner of the globe within reach by voice and visual communication, fairly and equitably available for use, proceed with all possible promptness."

—— Dr. Edward R. Sharp, Director Emeritus of the NASA Lewis Research Center since his retirement in January, died. Joining NACA in 1922, Dr. Sharp was the first Manager of the Lewis Flight Propulsion Laboratory in 1941, and became its Director in 1947. He had received the U.S. Medal for Merit from President Truman in 1947.

—— Joint FAA–DOD–NASA "Commercial Supersonic Transport Aircraft Report," prepared by a joint task force, said that the development of a commercial transport airplane to fly three times the speed of sound (mach 3) was feasible and could be developed by 1970–71.

July 25: NASA reported that one of the Tiros III cameras was inoperative, but that duplicate camera was producing high-quality pictures. Over 3,500 cloud cover pictures had been transmitted since the launching of Tiros III on July 12.

—— Titan ICBM with self-contained inertial guidance fired 5,000 miles down Atlantic Missile Range.

July 26: Cosmonaut Gagarin participated in anniversary celebration in Havana of July 26 revolutionary movement of Fidel Castro.

July 27: Third USAF Minuteman missile successfully flown on 4,000-mile flight down Atlantic Missile Range.

—— France announced plans to launch its first satellite by the end of 1964.

July 28: NASA and the American Telephone & Telegraph Co. signed a cooperative agreement for the development and testing of two, possibly four, active communication satellites during 1962. A.T. & T. would design and build the TSX satellites at its own expense, and would reimburse NASA for the cost of the launchings by Thor-Delta vehicles at Cape Canaveral. Relationship between this contract and the overall NASA communications satellite program aimed at early development of an operational system was explained at a NASA press conference.

—— NASA invited 12 companies to submit prime contractor proposals for the manned lunar Apollo spacecraft by October 9.

—— NASA representatives meeting with Arnold Engineering Development Center (AEDC) fixed the guidelines for the Centaur propulsion system testing program.

—— Interviewed in the Netherlands West Indies, en route to Brazil, Maj. Yuri Gagarin said that his next assignment would be a flight to the Moon. Asked about U.S. efforts, he reportedly stated that "there is a place on the Moon for everybody."

July 29: Chief of Japanese Weather Bureau, Kiyoo Wadachi, reported that 30 observations from Tiros III had been received from the United States.

—— World press reported opposition of astronomers to proposed USAF placement of 350 million needles into a 2,000-mile-high Earth orbit to test their feasibility as reflectors for global communications.

July 30: Draft text of program of the Soviet Communist Party to be presented to its 22d Congress in October was released in English by Tass, official Soviet press agency. This new program, the first proposed since the one submitted by Lenin and adopted in 1919, made no direct or indirect reference to space exploration. On the role of science, it stated: "The party will do everything to enhance the role of science in the building of Communist society, it will encourage research to discover new possibilities for the development of the productive forces, and the rapid and extensive application of the latest scientific and technical achievements, a decisive advancement in experimental work, including research directly at enterprises, and the efficient organization of scientific and technical information and of the whole system of studying and disseminating progressive Soviet and foreign methods. Science will itself in full measure become a productive force * * *."

July 31: NASA's Tiros II transmitted photograph of a major storm off the south tip of Africa. Launched on November 23, 1960, Tiros II was expected to only have a useful lifetime of about 3 months.

—— NASA awarded contract to University of Michigan to continue to provide research instrumentation for measurement of temperatures and winds at altitudes up to 150 kilometers with Nike-Cajun and other sounding rockets.

—— NASA provided for transfer of funds to ONR for balloons, launching services, and related expenses in connection with high-altitude measurements of electron, low-energy proton, and alpha-particle spectrum of primary cosmic radiation to be conducted by the University of Chicago from Uranium City, Saskatchewan, Canada.

July 31: At Cape Canaveral with the President's Missile Sites Labor Commission, Secretary of Labor Goldberg made public President Kennedy's message praising the voluntary, no-strike, no-lockout pledges covering labor-management relations at missile and space sites. The President's message stated that "the Nation cannot afford the luxury of avoidable delay in our missile and space program. Neither can we tolerate wasteful and expensive practices which add to the great financial burden our defense effort already places on us."

—— Atlas E fired from Atlantic Missile Range with simulated atomic fuel cores to demonstrate dispersal on reentry into the atmosphere of the radioactive material in an atomic space generator.

—— Vice Adm. T. G. W. Settle (Ret.) stated in Washington that Navy blimps should have been used in recovery of Mercury capsule, a proposal submitted to Navy 2 years ago, and which would have avoided recovery difficulties of Liberty Bell 7 and Astronaut Grissom. Settle pointed out that Navy had announced the end of its lighter-than-air program in June 1961.

During July: Langley Research Center simulated spacecraft flights at speeds of 8,200 to 8,700 feet per second in approaching the Moon's surface. With instruments preset to miss the Moon's surface by 40 to 80 miles, pilots with control of thrust and torques about all three axes of the craft were able to learn to establish orbits 10 to 90 miles above the surface, using a graph of vehicle rate of descent and circumferential velocity, an altimeter, and vehicle attitude and rate meters, as reported by M. J. Queijo and Donald R. Riley of Langley Research Center.

—— "Celestial simulator" at Jet Propulsion Laboratory in final checkout, an "instant universe" chamber which can duplicate white light and infrared point sources of solar system bodies likely to be used for navigation and attitude control of spacecraft.

—— U.S.S.R. has scheduled "at least two more manned space flights this year, one to circle the Earth, the other perhaps the Moon," according to Dr. Grigori A. Tokaty, head of Northhampton College of Advanced Technology, London, England. Former director of Russia's long-range rocket group, Tokaty also stated that the U.S.S.R. was planning to establish "one or two" unmanned lunar stations in 1962.

—— U.S.S.R. claimed three new world aircraft weight-lifting records for the Tu–114, in a flight from Vnukovo Airfield in which a 30,035-kilogram load was carried to an altitude of 41,125 feet, I. Sukhomlin as pilot.

AUGUST 1961

August 1: NASA directed Marshall Space Flight Center to enter contract negotiations with contractors for procurement of five operational Atlas-Centaur vehicles. These launchings were planned to begin in second quarter of 1964.

—— NASA Ranger I launch from Atlantic Missile Range postponed at T minus 15 minutes because of failure of ground-support equipment.

—— NASA Apollo briefing held at Space Task Group for all prime contractors interested in submitting bids.

August 2: NASA headquarters announced that it was making a worldwide study of possible launching sites for Moon vehicles; the size, power, noise, and possible hazards of Saturn-Nova type rockets requiring greater isolation for public safety than presently available.

—— Dr. Sydney Chapman of the British Royal Observatory reported at Langley Research Center-National Research Foundation-Virginia Polytechnic Institute conference that evidence suggested existence of a third radiation belt surrounding the Earth—except for areas above the poles—at altitudes between 20,000 and 28,000 miles.

—— USAF announced that two Lockheed U-2 aircraft would begin series of air-sampling flights from Okinawa.

August 3: USAF Discoverer XXVIII (total payload weight of 2,100 pounds) launched but did not attain orbit.

—— Tiros II transmitted photograph of a major storm in the Northwest Pacific Ocean.

August 5: Segmented solid-propellent rocket engine fired by United Technology Corp. at Sunnyvale, generating over 200,000 pounds of thrust in 80-second firing. Developed under NASA contract, center section of engine contained over 55,000 pounds of propellant, the largest single piece yet manufactured in the United States.

—— First Saturn (SA-1) booster began water trip to Cape Canaveral on Navy barge *Compromise* after overland detour around Wheeler Dam.

August 6: U.S.S.R. launched Vostok II into orbit carrying Maj. Gherman S. Titov. Spacecraft weighed 13 pounds more than Vostok I (April 12) and progress of Cosmonaut Titov's flight was reported continuously on Radio Moscow.

—— In press conference at Hyannis Port, Mass., U.S. Ambassador to the U.N. Adlai Stevenson, said: "Russia's scientific contribution to the conquest of outer space commands our admiration. Orbiting a new astronaut for a longer period of time is another step forward * * * this event [Vostok II] sharpens the need for some international action to regulate the use of outer space for peaceful purposes, and to keep the arms race from spreading to that field. The President has recently announced his proposal for cooperative sharing of communications and weather satellites. We hope the Russians won't delay longer in joining us in cooperation."

August 6: February report of the Space Science Board of the National Academy of Sciences was released recommending exploration of the Moon and planets "as the official goal of the U.S. space program and clearly announced, discussed, and supported."

August 7: Reported from Moscow that Major Titov has successfully landed in Vostok II after 17 orbits and 25 hours, 18 minutes, the first test of man's reaction to prolonged weightlessness. This was the second manned orbital flight, the first manned flight of more than one orbit.

―――― A joint message issued by Tass for the Soviet Party's Central Committee, the Cabinet, and the Presidium of the Supreme Soviet said: "Our achievements in the exploration of outer space are placed at the service of peace and scientific progress, for the benefit of all people on our planet."

―――― Two U.S. Air Force officers were sealed in space simulator for 17-day test of man's reaction to almost pure oxygen at 35,000 feet altitude, at School of Aerospace Medicine, Brooks Air Force Base, Tex. Emerging on August 25, Lts. B. Appel and J. Slider had eaten dehydrated food, drunk water processed from the atmosphere and their own body wastes, and were pronounced in good physical condition.

August 8: Over 100 foreign weather services were invited jointly by NASA and the U.S. Weather Bureau to participate in the Tiros III experiment for a 9-week period beginning today. The program provides cooperating services with an opportunity to conduct special meteorological observations synchronized with passes of the satellite.

―――― Atlas F successfully fired 5,000 miles from Atlantic Missile Range.

August 9: NASA selected MIT's Instrumentation Laboratory to develop the guidance-navigation system for Project Apollo spacecraft. This first major Apollo contract was required since guidance-navigation system is basic to overall Apollo mission. The Instrumentation Laboratory of MIT, a nonprofit organization headed by C. Stark Draper, has been involved in variety of guidance and navigation systems developments for 20 years.

―――― Enormous reception for Cosmonaut Gherman S. Titov in Red Square, Moscow. That evening at a Kremlin reception, Premier Nikita Khrushchev made an impromptu speech in which he asserted that the Soviet Union could construct a rocket with an explosive warhead equivalent to 100 million tons of TNT.

―――― Dr. Clifford C. Furnas, chancellor of the University of Buffalo, was appointed Chairman of the Defense Science Board by Secretary McNamara.

August 10: X–15 (No. 1) on its first flight with new XLR–99 engine was flown to 2,735 miles per hour by Comdr. Forrest S. Petersen, U.S. Navy, at Edwards Air Force Base.

―――― In regular press conference, President Kennedy stated that "we are spending as much money and devoting as large a percentage of scientific personnel, engineering, and all the rest, as we possibly can to the space program. We are constantly concerned with speeding it up. We are making what I consider to be a maximum effort."

―――― In passing NASA fiscal year 1962 appropriations, Congress cut $226,686,000 requested for salaries and expenses to $206,750,000.

August 11: Thomas F. Dixon of North American Aviation was appointed Director of NASA's Office of Launch Vehicle Programs (OLVP), effective September 18, 1961. He replaced Maj. Gen. Don R. Ostrander, U.S. Air Force, who returned to military duty as Vice Commander of AFBSD (AFSC), having served as first Director of OLVP since December 16, 1959.

—— Project West Ford received approval in National Aeronautics and Space Council policy statement released at the National Academy of Sciences by Presidential Scientific Adviser Jerome B. Wiesner. Conceived at MIT's Lincoln Laboratory, project proposed placement of 350-million copper threads (0.7-inch long and 0.001-inch diameter) into a 5-mile wide and 24-mile long belt around the Earth from a satellite, which would serve as reflector antennas for extremely short wave lengths (8,000 megacycles), perhaps expanding usable frequency channels.

—— Vostok II press conference held in Moscow, featuring President of the Soviet Academy of Sciences, Matislav Keldysh, and Cosmonaut Maj. Gherman S. Titov.

—— NASA announced negotiation of a contract with Hughes Aircraft for construction of three experimental synchronous communications satellites.

—— Aerojet-General Corp. announced first successful underwater launching of a liquid-fueled rocket, an Aerobee fired from a water test basin at Azuza, Calif.

—— NASA Langley Research Center awarded contract to Marquardt Corp. to increase structural wind tunnel testing temperature from 600° to 2,000° F.

August 12: Echo I completed first year in orbit, still clearly visible to the naked eye, after 4,480 orbits and traveling 138 million miles. Echo I provided basis for over 150 communications experiments, recent ones indicating only a 40-percent reduction in transmission reflection caused by the changed shape. Echo I provided significant data on atmospheric drag and solar pressure.

—— Aerobee 150-A fired with liquid hydrogen experiment from Wallops Island.

—— Record six Polaris missiles fired underwater in 1 day by U.S.S. *Abraham Lincoln*.

August 14: Navy barge *Compromise*, carrying first Saturn booster, stuck in the mud in the Indian River just south of Cape Canaveral. Released several hours later, the Saturn was delayed only 24 hours in its 2,200-mile journey from Huntsville.

—— Swedish scientists fired a U.S. Arcas rocket to 55.8-mile altitude from Arctic Circle test range at Vidsel.

August 15: Explorer XII (S-3) placed into highly eccentric orbit by Thor-Delta from Atlantic Missile Range, which would provide detailed evaluation of behavior of energetic particles between 170- and 50-000-mile altitude. Under Goddard Space Flight Center, this "windmill" satellite carried six experiments developed by Ames Research Center, State University of Iowa, University of New Hampshire, and Goddard Space Flight Center. Several days were required to confirm orbit.

August 15: Sir Bernard Lovell, director of the radiotelescope at Jodrell Bank, England, expressed concern in an interview about the USAF Project West Ford, to place 350 million small pieces of wire into an orbital band encircling the Earth at a height of 500 to 1,000 miles. Sir Bernard stated that "the published intention of the plan is to provide a reflector for radio communications from one side of the Earth to the other. Those of us who have studied this notice it is being carried out under the auspices of the USAF and not the NASA. * * *"

—— Dr. Frank B. Voris, captain, Medical Corps, U.S. Navy, Navy liaison officer to Project Mercury since 1958, reported aboard as Assistant Director for Aerospace Medicine in NASA's Office of Life Science Programs.

August 16: F-1 rocket engine tested in first of firing series of the complete flight system.

—— The International Academy of Astronautics, meeting in Paris, named Sir Bernard Lovell, Director of Jodrell Bank radiotelescope, as the first winner of the Daniel and Florence Guggenheim International Astronautics Award.

—— Centaur vehicle C-1 was tested and accepted by Marshall Space Flight Center.

August 16–18: General meeting of the International Astronomical Union at the University of California, Berkeley.

—— International Hypersonics Conference held at MIT.

August 17: NASA announced that Explorer XII had successfully completed first orbit, radioing data on magnetic fields and solar radiation from an apogee of near 54,000 miles and perigee within 170 miles of the Earth.

—— Tiros III spotted two storm cells about 500 miles south and southwest of Hawaii, reports which alerted Honolulu and Guam of these previously unknown potential typhoons.

—— President Kennedy signed into law the bill providing NASA appropriations for fiscal year 1962 of $1,671,750,000.

—— USAF Blue Scout launched from Atlantic Missile Range, radio contact lost during the fourth stage with payload intended to reach 140,000 miles into space.

August 18: NASA announced that analysis of Project Mercury suborbital data indicated that all objectives of that phase of the program had been achieved, and that no further Mercury-Redstone flights were planned.

—— Announced that NASA had decided to add 15 Agena B vehicles to the original Agena B program.

August 19: Controversy over Project West Ford aired at International Astronomical Union meeting at Berkeley, Calif.

August 21: NASA held a news conference on Explorer XII, at which the great amount of continuous coverage on interrelated data in its eccentric orbit was pointed out.

—— DC-8 jet airliner flown beyond mach 1 in experimental flight by Douglas Aircraft.

August 22: University of Michigan astronomers reported reception of natural radio signals from the planet Mercury.

—— Republic of China announced plans to initiate a rocket research program.

August 23: Ranger I test satellite of unmanned lunar spacecraft, launched from Atlantic Missile Range by Atlas D-Agena B into low parking orbit, but did not attain its programed eccentric orbit.

——— Maj. Gherman S. Titov, in his serialized account of his orbital flight in Pravda, described the state of weightlessness.

——— Saturn H-1 engine drop-tested into salt water at Cape Canaveral, then returned to Marshall Space Flight Center for inspecting, cleaning, and static firing.

August 24: NASA announced decision to launch manned lunar flights and other missions requiring Saturn and Nova class vehicles from expanded Cape Canaveral facilities. Based upon national space goals announced by the President in May, NASA plans call for acquisition of 80,000 acres north and west of AFMTC, to be administered by the USAF as agent for NASA and as a part of the Atlantic Missile Range. Decision followed intensive NASA-DOD survey for launching facilities, including trajectory advantages, overflight or booster impact hazards, air and water transportation, instrumentation support, and cost, time, and land availability advantages. Expansion of Cape Canaveral was noted as first of three major steps in accelerating the U.S. space program, the remaining two steps being a manned space flight research center, and a booster fabrication and test facility.

——— Mercury-Atlas 4 launch postponed.

——— NORAD charts showed that flight of Vostok II was tracked continuously.

August 25: Explorer XIII (S-55A) placed into orbit by NASA Scout from Wallops, a micrometeorite counting satellite developed by Langley Research Center and Goddard Space Flight Center.

——— NASA announced selection of Blaw Knox Co. to conduct second-phase feasibility study for a 240-foot diameter deep space tracking antenna for Jet Propulsion Laboratory's deep space instrumentation facility at Goldstone, Calif.

August 26: Explorer XIII, popularly referred to as the "beer can satellite" because of its micrometeorite counting structure, completed its 15th orbit.

——— Aerojet-General fired largest solid-fuel-rocket motor to date, over one-half million pounds of thrust, at Sacramento, Calif. This motor weighed over 70 tons and was made in several segments which were joined together at the static test site.

——— Northrop T-38 (Talon) jet flown 842.6 miles per hour to claimed world speed record for women, by Jacqueline Cochran at Edwards Air Force Base, Calif.

August 27: In a letter to the President, 35 Members of Congress urged that a decision on the Nation's satellite communication system be delayed to determine "whether such a system should be publicly or privately owned and under what circumstances."

——— The Soviet Communist Party organ, Pravda, explained why Russian space techniques and the names of spaceship designers were kept secret as follows: "A corrupt capitalist society, by its very nature, is extremely capable of turning the greatest peaceful achievements of mankind into the total means of destruction of mankind. This is why it is risky to open even the smallest loopholes in the world of Soviet rocket technique for the gentlemen who are lagging considerably behind as far as their technique is concerned, but who become militarily agitated and distracted from an honest program of general and complete disarmament and who mumble something about the right of inspection of neighbors' orchards and storerooms. That is why the wonderful group of heroes who insured the mastering of the cosmos remain nameless until now."

August 28: NASA selected Vitro Engineering Co. for negotiation of a design contract for an engine maintenance and disassembly building, one of the facilities to be a part of the National Nuclear Rocket Development Center.

—— Lt. Gen. Bernard A. Schriever, Commander of Air Force Systems Command, said that plans to orbit a monkey in Discoverer XXII (March 30) were canceled at the last minute, and that such plans had not been rescheduled.

—— Reported that Martin Co. originally required 75,000 man-hours to produce the first Titan I's, which had now been reduced to 19,000 man-hours per Titan. Man-hour rate for the first five Titan II's averaged 35,000 man-hours on each one.

—— Lt. Hunt Hardisty, U.S. Navy, pilot, and Lt. Earl H. DeEsch flew an F4H Phantom II over the 3-kilometer course of Holloman Air Force Base, N. Mex., and averaged 902.769 miles per hour for a new world's record for speed at low altitude.

August 29: NASA announced that Explorer XIII launched on August 25 had reentered the atmosphere. Transmitting considerable data on micrometeoroids, spacecraft was last heard on August 27 by the Minitrack facility at Antofagasta, Chile.

—— NASA Associate Administrator Seamans announced the addition of four additional Ranger spacecraft, bringing the total to nine, the number of Rangers to be launched in this phase of the lunar exploration program. "This new third phase of the Ranger program is a part of the general acceleration of the program to land an American on the Moon by 1970," said Dr. Seamans.

—— Ranger I completed 100 orbits, transmitting data on all engineering devices and eight scientific experiments. It was expected to come down soon.

August 30: USAF Discoverer XXIX launched into polar orbit with 300-pound data capsule, from Pacific Missile Range.

—— NASA announced that Ranger I spacecraft had reentered the Earth's atmosphere. Launched on August 23, Ranger I made 111 orbits, traveled almost 3 million miles, and its orientation, communications, and electronic systems performed satisfactorily.

—— USAF Minuteman ICBM exploded seconds after firing in silo at Cape Canaveral due to guidance malfunction.

August 31: U.S.S.R. announced policy of resumption of nuclear weapon testing which had been suspended March 31, 1958, and that bombs can be delivered anywhere in the world by "powerful rockets like those Majs. Yuri Gagarin and Gherman Titov rode to begin their unrivaled space flights around the Earth."

During August: NASA site selection team headed by John F. Parsons, Associate Director of Ames Research Center, toured possible sites for a manned spacecraft center.

—— New wind tunnel became operational at Ames Research Center, capable of research on reentry problems at speeds of mach 7.5, 10, and 15.

—— With successful launch of Explorer XII on August 15, NASA Delta launch vehicles had successfully launched five satellites out of six attempts, the only failure being the first attempt. Delta's high reliability record began with Echo I on August 12, 1960, and includes Tiros II and III, and Explorers X and XII. Built by prime contractor Douglas Aircraft, the NASA Delta launch vehicle consists of a Thor first stage (Rocketdyne MB-3 liquid engine), Aerojet-General second stage (AJ-10-118, an improved Vanguard second stage), and an ABL third stage (X-248 spin-stabilized version of Vanguard third stage).

—— New 210-foot diameter radiotelescope began operations at Parkes, New South Wales, operated by a group of scientists headed by Dr. E. G. Bowen, radar pioneer.

—— NASA-DOD Large Launch Vehicle Planning Group in session since August 1 to study the policy, management structures, and requirements of launch vehicles beyond the size of Saturn. Meeting in NASA headquarters, the group was headed by Dr. Nicholas E. Golovin, technical assistant to the Associate Administrator of NASA.

—— Announced that RCA scientists determined the distance between Earth and Venus to an accuracy of 200 miles using 84-foot tracking antenna.

SEPTEMBER 1961

September 1: White House announced that the U.S.S.R. had resumed testing of nuclear weapons early this morning, the first known nuclear test by U.S.S.R., United States, or Britain since the fall of 1958.

—— Per NASA management instructions, all space vehicles and spacecraft under cognizance of NASA were to be equipped with fail-safe devices for terminating electromagnetic transmissions at the completion of their planned useful life.

—— Three parachuting skindivers recovered capsule of Discoverer XXIX, the seventh recovery of an object from orbit in the USAF Discoverer program. Capsule had made 33 orbits and contained human, animal, and soil life samples.

—— NASA Administrator Webb, appearing before the Senate Committee on Aeronautical and Space Sciences, requested $60 million additional for fiscal year 1962 for the acquisition of 80,000 acres adjoining Cape Canaveral for launching facilities for the expanded space program.

—— NASA issued its "Program Evaluation and Review Technique Handbook," its adaptation of the Navy PERT program management system.

September 2: Scientists at Nagoya University, Japan, were reported to be training monkey for space flight next year, hopefully in conjunction with Japanese Government-financed rocket program carried out by Tokyo University's Institute of Industrial Science.

September 3: Thirty days' exposure to simulated vacuum of space killed bacteria by causing them to disintegrate molecule by molecule, was finding of studies reported by the Materials Testing Laboratory of Hughes Aircraft Co. Dr. Charles G. Walence reported that sterilization of space vehicles probably could be eliminated from current planning.

September 5: Authorization for NASA to acquire necessary land for additional launch facilities at Cape Canaveral was approved by the Senate.

—— In an interview with C. L. Sulzberger, Premier Khrushchev reviewed the world crisis in detail. "In a strange Darwinian interpretation of the advance of nations," Sulzberger reported, "Mr. Khrushchev jokingly considers that the United States is still in the stage of 'jumping' while the Soviet Union has learned how to 'fly.' This refers to the Earth-orbiting successes of the Soviet spacemen, Maj. Yuri A. Gagarin and Maj. Gherman S. Titov."

September 5-7: International Conference on Science and World Affairs held at Stowe, Vt., at which nonofficial proposals for a joint United States-U.S.S.R. space program were considered by delegates of 12 nations including the United States and the Soviet Union.

September 6: USAF Titan successfully launched from Atlantic Missile Range, making 6,100-mile flight.
— AEC announced that U.S.S.R. had detonated a fourth nuclear device in the atmosphere, at a site east of Stalingrad.
— After a series of six static firings, the Saturn SA-T2 booster was removed from the static test tower at Marshall Space Flight Center.

September 7: NASA announced that Government-owned Michoud Ordnance Plant near New Orleans would be the site for fabrication and assembly of the first stage of Saturn as well as for making stages for larger booster.
— Balloon flights to measure loss of radiation from the Van Allen radiation belts—the "dumping profile" experiment—was announced by National Science Foundation. Flights are part of joint project by University of Minnesota and University of California, taking place along line from Flin Flon, Manitoba, to Waterloo, Iowa.
— USAF Titan with inertial guidance system successfully launched from Atlantic Missile Range, the second in as many days, impacting into target area over 5,000 miles down range.
— The Agena B vehicle 6002 was delivered to Atlantic Missile Range, in preparation for the Ranger 2 launch.

September 8: Deep space tracking antenna dedicated by United States and South African officials. Located 40 miles from Johannesburg, the antenna has operated since July in collaboration with Goldstone and Woomera, and tracked Ranger I.
— Reported from Stowe, Vt., that unofficial American-Soviet discussions on cooperative space exploration were near agreement, and that internationalized status for space similar to that achieved by treaty for the continent of Antarctica was under consideration. Delegations included seven members of President Kennedy's Science Advisory Committee and six members of the ruling body of the Soviet Academy of Sciences. The Stowe Conference was sponsored by the National Academy of Arts and Sciences of Boston, with costs defrayed by the Ford Foundation.

September 10: Tiros III photographed Hurricane Esther in process of formation, 2 days before hurricane-hunter aircraft verified winds of hurricane force. Quality of Tiros III pictures processed in 8 hours through the National Meteorological Center, Suitland, Md., was not alone sufficient for identification of a hurricane.
— On the same day, Tiros III also observed one-eighth of the Earth, providing data on two other hurricanes (Carla and Debbie), one dissipating hurricane (Betsy), two typhoons (Nancy and Pamela), and at least one other vortex storm.
— White House released "Project Horizon" report of task force created in March to establish goals to maintain America's primacy in aeronautics. The report made a strong recommendation for the development of a supersonic transport, among other recommendations.
— U.S.S.R. announced that it would launch a series of "more powerful and improved rockets" into the Central Pacific in tests between September 13 and October 15.

September 11: NASA selected North American Aviation to develop an upper stage (S-II) for an Advanced Saturn launch vehicle to be used on both manned and unmanned missions.

September 12: In a speech before the National Press Club, NASA Administrator Webb reviewed NASA's program and outlined the interest in the rendezvous-in-space technique for staging flights to the Moon and nearby planets.

—— X-15 flown to record 3,614 miles per hour by NASA's Joseph A. Walker at Edwards Air Force Base.

—— USAF Discoverer XXX launched into polar orbit from Pacific Missile Range.

September 13: Unmanned Mercury spacecraft orbited by Mercury-Atlas launch vehicle from Atlantic Missile Range and recovered 1 hour and 22 minutes after landing by destroyer U.S.S. *Decatur*. This MA-4 (capsule 8) flight demonstrated, said NASA Space Task Group Director Robert Gilruth, that "Atlas has the capability to fly a man in orbit; it brought in for the first time the Mercury worldwide tracking network; and demonstrated the ability of the capsule and its systems to operate completely unattended."

—— Two experiments to measure atmospheric winds, temperature, and density in relatively high altitudes conducted from Wallops Island in two four-stage Argo D-4 rocket launches. Sodium clouds were released at near 120 statute miles and again at 228 miles in first launch, and at 118 and 230 miles in the second launch. French scientists participated by using special optical instruments to observe the brilliant orange and yellow clouds which stirred a rash of public inquiries to newspapers from hundreds of miles around.

—— U.S.S.R. announced that it had fired new, powerful carrier rocket more than 7,400 miles to within less than five-eighths of a mile from its Central Pacific target.

—— Soviet Marshal Kiril S. Moskalenko, chief of rocket forces, declared that 95 percent of all Soviet rockets fired reached their targets. (Article in Red Star timed to coincide with first firing of new rocket series in the Pacific.)

September 14: USAF C-130B cargo plane snagged the parachuting capsule of Discoverer XXX north of Hawaii, Capt. W. C. Schmensted as pilot.

—— White House released its reply to letter of August 27 signed by 35 Members of Congress which expressed concern over the private ownership of an operational communications satellite system. The White House memorandum stated that "any decisions as to control should preserve as much flexibility as possible," and reemphasized the administration policies including "maximum competition" in any system of private ownership.

—— Resolutions calling for the creation of an international space year program and an international space agency, both under the auspices of the United Nations, introduced in the Senate by Senator Hubert Humphrey.

September 14: AEC announced that the Soviet Union had fired its 10th nuclear blast in the current test series begun 2 weeks ago.

September 15: White House announced that AEC–DOD had conducted first U.S. nuclear weapons test since October 1958, an underground weapons development test at the Nevada testing site, one of low yield and which produced no fallout.

—— Army Nike-Zeus fired from White Sands Proving Ground met all test objectives, including controlled high-velocity in the atmosphere and evaluation of solid-fuel rocket motors and guidance system.

—— Marshall Space Flight Center's Procurement and Contracts Office reported that a contract was let to the Noble Co. for disassembling the Redstone gantry at Atlantic Missile Range and reassembling and erecting it at Pacific Missile Range. This gantry would be erected on pad 75-1-1 at Vandenberg Air Force Base for use with Thor-Agena B launches after January 1962.

September 16: Congressman Overton Brooks, of Louisiana, chairman of the House Committee on Science and Astronautics since its creation in January 1959, died at Bethesda Naval Hospital.

September 17: USAF Discoverer XXXI placed into polar orbit from Pacific Missile Range by Thor-Agena.

—— Soviet Union announced that 12 altitude and speed records had been broken by its twin-jet M-10 antisubmarine seaplane.

September 18: First of four scheduled Skylark rocket firings was launched from Woomera in the joint United States-Australian ultraviolet survey of the southern skies.

September 19: NASA Administrator Webb announced that location of the new Manned Spacecraft Center would be in Houston, Tex., the conclusion of an intensive nationwide study by a site selection team. The Manned Spacecraft Center would be the command center for the manned lunar landing mission and all follow-on manned space flight missions. This announcement was the third basic decision on major facilities required for the expanded U.S. Range and the establishment of the spacecraft fabrication center at the Michoud Ordnance Plant near New Orleans, La.
Center at the Michoud Ordnance Plant near New Orleans, La.

—— Recovery of capsule of Discoverer XXXI was called off as capsule and satellite (launched Sept. 17, 1961) failed to separate and both remained in orbit.

—— USAF Bomarc B launched from Eglin Air Force Base, Fla., on command from SAGE Center at Montgomery, Ala., destroyed supersonic Regulus launched from Venice, Fla.

—— Air Force Systems Command formed a Bioastronautics Division, effective October 1, to consolidate all USAF applied research in this area into a single organization. School of Aerospace Medicine, now under Air Training Command, becomes a part of Air Force Systems Command.

September 19: In a speech to the USAF Worldwide Information Conference at Philadelphia, Maj. Gen. Daniel E. Hooks, Commander of the Office of Aerospace Research, reported that predictions of OAR's Solar Laboratory at Sacramento Peak, N. Mex., were borne out by the flights of the U.S.S.R.'s Vostok I and II. High proton shower activity associated with solar flares had been predicted for April 1961, except from April 11 through 14 (Major Gagarin's flight in Vostok I was on April 12). August 6, the day of the launching of Major Titov's 17-orbit flight, was the "safest day" for low solar activity on record since 1955.

September 20: Robert Gilruth and other officials of the Space Task Group made survey of the new site of the Manned Spacecraft Center near Houston, Tex., to seek temporary operational quarters as soon as possible. Permanent quarters will be constructed under the supervision of the Army Corps of Engineers.

September 21: D. Brainerd Holmes was appointed NASA's Director of Manned Space Flight Programs. As general manager of RCA's Major Defense Systems Division, Holmes was project manager for the ballistic missile early warning system (BMEWS).

—— Representative George P. Miller, Democrat, of California, was named chairman of the House Science and Astronautics Committee.

—— Soviet Union protested to the U.S. National Academy of Sciences that Project West Ford might endanger Soviet cosmonauts, protest contained in a letter to Detlev W. Bronk, president of the NAS, signed by Matislav Keldysh, president of the Soviet Academy of Sciences.

September 22: Announced at Space Task Group that a 30-cubic-foot balloon would be installed in Mercury spacecraft to allow for ship recovery should helicopter be forced to drop it as happened during the MR-4 recovery.

September 23: U.S.S.R. announced that third flight of current series of Pacific tests of new multistage carrier rocket was successful.

—— NASA planned to spend $6 million on expansion of its Langley facilities in this fiscal year, according to Representative Thomas N. Downing after a conversation with NASA Administrator James E. Webb. Representative Downing said he was satisfied that the lower peninsula area of Virginia would not suffer financially when the Space Task Group moved to Houston, Tex.

September 24: Administrator Webb announced major organizational changes and top-level appointments in NASA. Keyed to the Nation's accelerated civilian space program, changes provided clearer focus on major programs, and provided center directors with a louder voice in policymaking and program decisions. Effective November 1, major headquarters program offices would be headed as follows: Ira H. Abbott, Director of the Office of Advanced Research and Technology; Homer E. Newell, Director of the Office of Space Sciences; D. Brainerd Holmes, Director of the Office of Manned Space Flight; and an Office of Applications Programs with no director yet named. Thomas F. Dixon was appointed Deputy Associate Administrator. Abe Silverstein was named Director of the Lewis Research Center and Robert R. Gilruth, Director of the Manned Spacecraft Center.

September 24: Speaking at the Air Force Association convention in Philadelphia, Gen. Bernard A. Schriever, Commander of the Air Force Systems Command, said that "the United States has been notably slow to recognize the military application of new inventions. Two of the most significant inventions of this century—the airplane and the liquid-fuel rocket—are American inventions. Yet in each case their first application was made by other nations." Gen. Schriever also stated: "We should recognize that there is no inherent difference between basic military and nonmilitary space technology. What really matters is not the technology but the intent * * * space power must become a vital part of our national strength and security."

—— NASA made a grant to Stanford University's School of Medicine for development of design of payload instrumentation to determine existence of life forms on nearby planets, a project under the direction of Dr. Joshua Lederberg and Dr. Elliott C. Leventhal.

September 25: In a stirring address to the Assembly of the United Nations meeting in New York, President John F. Kennedy called for "world law in the age of self determination" rather than a "world war in the age of mass extermination." Among basic proposals for waging effective peace, he urged "keeping nuclear weapons from seeding new battlegrounds in outer space." In projecting the theme that "the events and decisions of the next 10 months may well decide the fate of man of the next 10,000 years," President Kennedy spoke of the impact of space exploration as follows:

"As we extend the rule of law on Earth, so must we also extend it to man's new domain—outer space.

"All of us salute the brave cosmonauts of the Soviet Union. The new horizons of outer space must not be riven by the old bitter concepts of imperialism and sovereign claims. The cold reaches of the universe must not become the new arena of an even colder war.

"To this end, we shall urge proposals extending the United Nations Charter to the limits of man's exploration in the universe, reserving outer space for peaceful use, prohibiting weapons of mass destruction in space or on celestial bodies, and opening the mysteries and benefits of space to every nation * * *."

September 25: Dr. George N. Constan of Marshall Space Flight Center named as acting manager of the new NASA Saturn fabrication plant near New Orleans by Director von Braun of Marshall Space Flight Center.

September 26: NASA bidders conference on a contract to produce the booster (S-I) stage of the Saturn vehicle was held at the Municipal Auditorium, New Orleans.

—— Meeting to examine the modification of C-133 aircraft for carrying Saturn S-IV stages was held at NASA Langley Research Center.

September 27: Ion engine developed on NASA contract demonstrated at Hughes Aircraft Laboratories.

September 28: NASA announced that instrumented Venus probe to be launched next year would be launched by an Atlas-Agena B rather than a Centaur rocket as originally planned.

—— Solar flare studied by Explorer XII and Injun I, readings within a few hours indicated energetic protons with velocities of near 10,000 kilometers per second, while 2 days after the solar flare both satellites saw a sudden increase in the intensity of low-energy protons (c. 10 mev) concurrently with a magnetic storm on Earth and bright aurora at low altitude. It was concluded that low-energy protons traveled slowly from the Sun with the magnetic stormcloud.

—— X-15 (No. 2) flown to 100,800 feet and 3,600 miles per hour by Comdr. Forrest S. Petersen, U.S. Navy, at Edwards, Calif.

September 28-29: Pair of spinup rockets on Tiros II successfully fired after more than 10 months in orbit.

September 29: USAF awarded three contracts for speeding development of the Dyna-Soar, a manned orbital space glider. Receiving contracts were Boeing Co. for development of the glider and related systems, Radio Corp. of America for communications and tracking devices, and Minneapolis-Honeywell Regulator Co. for the guidance system.

—— Dr. Charles M. Herzfeld, of the National Bureau of Standards, joined DOD's Advanced Research Projects Agency to coordinate the Project Defender program.

—— Navy Polaris (A-3) with modified second stage testing freon gas regulator launched from pad at Atlantic Missile Range.

September 30: NASA Office for the United Nations Conference headed by Dr. John P. Hagen was closed, effective this date. Continued uncertainty of arrangements for a conference on the peaceful uses of outer space within the United Nations made such a move necessary. Sustained NASA responsibility in this regard was assigned to the Office of International Programs.

—— Air Force Systems Command announced consolidation of all USAF research and development in bioastronautics under single management. The new Bioastronautics Division, Air Force Systems Command, would have its headquarters at Brooks Air Force Base, San Antonio, Tex.

—— Two X-15 test pilots, USAF's Maj. Robert M. White and NASA's Joseph A. Walker, were jointly awarded the 1960 Iven C. Kincheloe Memorial Award as the Nation's outstanding test pilots; award of the Society of Experimental Test Pilots.

September 30: NASA received an additional $10 million for salaries in a supplemental appropriation bill approved by Congress, enough to hire an additional 1,250 persons.

During September: National Bureau of Standards and the Instituto Geofisico de Huancago of Peru initiated construction of the Jucamarca Observatory, a 6-million-watt pulse transmitter and a 22-acre antenna with 9,216 crossed dipoles mounted above a reflecting screen. Located 17 miles east of Lima, Peru, the Observatory will be used for ground-based exploration of the upper atmosphere and space.

—— USAF established a Council of Scientists to be comprised of senior civilian scientists of major Air Force organizations, Dr. Leonard S. Sheingold as Chairman.

—— Congress appropriated funds to the U.S. Weather Bureau for implementation of the National Operational Meteorological Satellite System. To phase in as early as technology warrants and to continue expanding the operational capability through the early Nimbus launchings by NASA, the system planned to be fully operational by 1966 as Nimbus system became operational. The system would include data acquisition stations in northern latitudes, communications for transmitting the data, and a National Meteorological Center to receive, process, analyze, and disseminate the derived information over domestic and international weather circuits.

OCTOBER 1961

October 2–7: Twelfth Congress of the International Astronautical Federation held in Washington, D.C.

October 2: NASA Deputy Administrator Dryden and Soviet Academy of Sciences official Dr. Leonid I. Sedov both appealed for greater international cooperation and exchange of information in the peaceful exploration of space in their speeches at the opening of the 12th World Congress of the International Astronautical Federation.

—— NASA conducted a press conference for foreign correspondents attending the IAF Congress, pointing out that some 40 nations are now participating in NASA programs or are obtaining NASA help for their respective space programs. Director of the Office of International Programs, Arnold Frutkin, pointed out that growing space research cooperation would soon include a university training program in which 100 foreign students would work at American universities on peaceful space experiments.

—— USAF Atlas E missile made successful 5,000-mile flight at Atlantic Missile Range. The payload included the guidance equipment for the Centaur rocket, radiation sensors, and a nose cone intended for the Minuteman. Data capsule was recovered.

October 3: House Science and Astronautics Committee released interim report on "Research and Development in Aeronautics," which concluded that "the welfare of the Nation, in both its economic and security aspects, is dependent in no small degree on continuing aeronautical research of high caliber."

—— Vice President Lyndon B. Johnson began tour of west coast missile and space installations.

—— Inhouse procurement policies and practices of NASA reviewed by headquarters and field personnel in conference at Lewis Research Center.

—— First regular meeting of the International Academy of Astronautics held in conjunction with the 12th Annual Congress of the International Astronautical Federation in Washington.

—— Dr. Vladimir A. Kotelnikov, of the U.S.S.R. Academy of Sciences, told the IAF that Russian radar returns from Venus indicated a value of 149,599,500 kilometers for the astronomical unit. This was a major revision of the value of 149,469,500 kilometers first released by the Russians in March. The Russian value compared with 149,598,820 kilometers obtained by Jet Propulsion Laboratory and the figure of 149,597,850 kilometers obtained by Lincoln Laboratory. Disagreement remains on whether Venus day is a 9-to-11-earth-day period or 225-earth-day period.

October 3–5: PERT (program evaluation and review technique) symposium held at Huntsville, Ala., sponsored by the American Institute of Industrial Engineering and the University of Alabama.

October 4: Beginning of the fifth year of the "space age," being the anniversary of the launching of Sputnik I (1957).

———— Project West Ford given final approval by the White House.

———— Maj. Robert Rushworth, U.S. Air Force, flew X-15 to 2,820 miles per hour, with bottom tailfin missing in programed malfunctions for test of stability and control.

———— State Department ruled that Soviet space scientists would not be allowed to visit the American Rocket Society's Space Flight Report to the Nation in New York on October 9–14, a reciprocal action prompted by Soviet restrictions on American scientists in the U.S.S.R.

———— Soviet scientists in Washington for 12th IAF Congress revealed that Maj. Gherman Titov was ill during his 17 orbits in Vostok II on August 6. Disorientation, nausea, and irregular heartbeat resulted from prolonged weightlessness, according to O. G. Gazenko and V. J. Yazdovsky of the Soviet Academy of Sciences.

October 5: USAF Atlas fired 9,000 miles from Atlantic Missile Range into Indian Ocean, carrying dummy nuclear warhead and a data capsule which was recovered.

October 5–6: Atomic Energy Commission announced that the Tory II-A-1 nuclear test reactor underwent tests on September 28 and October 5–6. Power levels of the test were not disclosed but advanced plans called for tests at full power. The test was for "about 1 minute" at temperatures "in excess of 2,000° F." and emission of radiation was "negligible."

October 6: USAF Titan I launched from Cape Canaveral carrying Titan II guidance system.

October 7: Soviet E-166 jet fighter flown to 1,482.039 miles per hour in closed 100-kilometer course, according to Moscow claim.

———— Second stage of Nike-Zeus exploded at 2-mile altitude in test launch.

———— U.S.S.R. launched fourth multistage rocket 7,500 miles into the Pacific.

October 8: Plans for a worldwide scientific study of the Sun, to begin in 1964 and continue for 18 months, were approved by scientists from 51 nations gathered in London for the triennial meeting of the International Council of Scientific Unions. Final plans to be drawn in Paris in April 1962 at a meeting of the International Committee on Geophysics, successor to the IGY.

———— In article in New York Times, Dr. Edward C. Welsh, Executive Secretary, National Aeronautics and Space Council, said: "In my view, we [the United States] do not have a division between peaceful and nonpeaceful objectives for space. Rather, we have space missions to help keep the peace and space missions to enable us to live better in peace."

———— In article in New York Times, Mr. George J. Feldman, consultant to the House Committee on Science and Astronautics, outlined several areas of international space law that urgently require solution, including sovereignty in space, liability for damage from spacecraft, conflicts of interest arising from space experiments, sovereignty claims on celestial bodies, and the international allocation of satellite radio frequencies. Communications satellites make latter point particularly critical, as well as posing an unprecedented problem in American contract and antitrust law.

October 9–15: American Rocket Society's 16th annual meeting and Space Flight Report to the Nation held in New York City.

October 10: NASA Argo D-4 rocket was launched from Wallops, reaching an altitude of 585 miles and landing 817 miles out in the Atlantic, to gather data on the density of electrically charged helium atoms in the upper atmosphere.

October 11: X-15 flown more than 40 miles into space—217,000 feet—and reached a speed of 3,647 miles per hour, Maj. Robert White, U.S. Air Force, as pilot. This was above 99.9 percent of Earth's atmosphere; pilot's heartbeat and respiration rose to twice above normal; and outside skin temperature of the X-15 rose to 900° F. on reentry.

—— NASA Administrator Webb, speaking to the American Rocket Society, said NASA scientists "are going to consider the rendezvous technique with great care before going ahead with Nova." Decision on whether to give priority to the rendezvous technique would probably be made by the end of 1961, he said.

—— Final report of House Committee on Science and Astronautics relating to their hearings on "Commercial Applications of Space Communications Systems" released, having among its conclusions:

(1) Because of worldwide interest and potential usefulness of a space communications system, the U.S. Government must "retain maximum flexibility regarding the central question of ownership and operation of the system."

(2) NASA will not only evaluate the various commercial proposals but will "conduct all space launches and retain direct control over all launching equipment, facilities, and personnel."

(3) Research and development of military space communications systems should continue to be conducted by DOD but all research and development in space communications "should be conducted under the general supervision of NASA in accordance with its statutory mandate to 'plan, direct, and conduct aeronautical and space activities'" as well as evaluate the technical merits of proposed systems.

—— In a luncheon address to the American Rocket Society, Gen. Bernard A. Schriever said: "I have been, am being, and, if the situation is not changed, will continue to be inhibited if our space efforts continue to be carried out under an unnecessary, self-imposed national restriction; namely, the artificial division between space for peaceful purposes and space for military purposes." Asserting USAF management experience in space systems, General Schriever added: "There is no short cut to the creation of a team of dedicated and experienced men with a tradition of accomplishment."

—— Jacqueline Cochran set woman's altitude record of 56,071.3 feet, in Northrop T-38 jet trainer at Edwards Air Force Base.

October 13: In speech at the American Rocket Society, Vice President Lyndon B. Johnson stated: "If I could get one message to you it would be this: The future of this country and the welfare of the free world depend upon our success in space. There is no room in this country for any but a fully cooperative, urgently motivated all-out effort toward space leadership. No one person, no one company, no one Government agency, has a monopoly on the competence, the missions, or the requirements for the space program. It is and it must continue to be a national job."

October 13: Discoverer XXXII was placed into polar orbit; its capsule contained components of USAF satellite systems. This marked the 100th successful firing of the Thor booster rocket.

—— The Ad Hoc Carrier Committee established by the FCC to make an industry proposal on the development and operation of commercial communications satellites recommended a nonprofit corporation be formed, to be owned by companies engaged in international communications, with the U.S. Government having one more representative on the board of directors than any one company. Western Union filed a minority statement proposing a public stock company arrangement to prevent dominance of the corporation by any one company.

—— The American Rocket Society presented its major annual awards as follows: Dr. Robert H. Goddard Memorial Medal to Dr. Wernher von Braun, Director of NASA Marshall Space Flight Center; Astronautics Medal to Comdr. Alan B. Shepard, Mercury astronaut, for his MR-3 flight of May 5; James H. Wyld Memorial Medal to Harrison A. Storms, Jr., of North American Aviation; Propulsion Medal to Robert B. Young of Aerojet-General Corp. for his role in development of Titan II engine; G. Edward Pendray Award to Kraft Ehricke for his contribution to astronautical literature; and Research Medal to Dr. James Van Allen of State University of Iowa for basic research.

—— On its second birthday in space, Explorer VII was still transmitting, although it had been scheduled to stop a year ago.

—— The Soviet Union announced it had fired fifth multistage rocket 7,500 miles into the Central Pacific, with all stages functioning perfectly and with the nose cone landing in the target area with a high degree of accuracy.

October 14: Capsule of Discoverer XXXII recovered by C-130 piloted by Capt. Warren Schensted, U.S. Air Force, the sixth aerial recovery of an ejected satellite capsule and Schensted's second catch. Capsule contained test objects including seed corn.

—— NASA Argo D-4 launched from Wallops Station carried United States-Canadian topside sounding satellite payload to 560-mile altitude.

—— U.S.S.R.'s Tass announced that the "Air Force Herald" would be retitled "Aviation and Cosmonautics" (Aviatsiga I Kosmonavtika), beginning in January 1962.

—— U.S.S.R. claimed a new world speed record for vertiplanes on a closed 62-mile circuit at 209 miles per hour. Tass said this exceeded the previous record of 191 miles per hour held by a New Zealander, G. Ellith, flying a British Rotordyne. The following day, Tass claimed a horizontal speed of 228 miles per hour for the Kamov vertiplane.

October 14-15: Sky Shield II provided aerospace control exercise for NORAD and SAC, including grounding of all commercial aircraft for 12 hours.

October 17: USAF-USN-NASA X-15 flown to 108,600 feet and a record speed of 3,900 miles per hour, piloted by Joseph Walker at Edwards, Calif.

October 18: NASA Scout fired payload to 4,261-mile altitude, obtaining data on the ionosphere.

October 18: First U.S. showing of films of Vostok II space flight by Gherman Titov, before the Maryland Academy of Sciences in Baltimore, was canceled at the last minute by a Soviet Embassy official. Film had been shown to press correspondents in Moscow on October 9.

———— James A. Van Allen was awarded the Elliott Cresson Medal of the Franklin Institute, Philadelphia, for "his many contributions and pioneering achievements in the field of space science. * * *"

October 18–22: The 20th American Assembly sponsored by Columbia University met to study the problems of space exploration and, in its report, recommended a proper balance with it and other programs in the national interest.

October 19: In a speech at Naval Research Laboratory, Harold Brown, Director of Defense Research and Engineering, said Government labs would hereafter be the "primary means" for carrying out military weapon programs; that DOD would seek an increase in the number of supergrade scientific positions and would ask for the same top pay for scientists as NASA has; that labs would be given increased status in the chain of command; and that lab directors will be given funds they can spend for research without prior approval.

———— NASA Administrator Webb, speaking at the 20th American Assembly, said the accelerated space program was necessary or else "we would see the Russians, with the advantage of their advance position in booster thrust, stay continuously ahead. * * * The cost over the 10 years of the accelerated program will very probably be less than if it were stretched out over 15 years."

———— NASA Scout launched from Wallops Island, Va., and placed 94-pound P–21 payload to 4,261-mile altitude in a study of the ionosphere.

October 20: Ranger test postponed at Atlantic Missile Range.

October 21: USAF Midas IV launched into polar orbit from Pacific Missile Range, and also carried Project West Ford payload.

October 22: NASA announced that Dr. Hiden T. Cox, executive director of the American Institute of Biological Sciences, would become Assistant Administrator for Public Affairs, and "charged with developing NASA policies to insure that the character, the intent, and the results of America's space effort are correctly and adequately interpreted to the people of this country and the world."

———— Assistant Secretary of State for International Organization Affairs, Harlan Cleveland, outlined in a speech at St. Louis University the seven-point program that the United States will propose to the United Nations General Assembly for guaranteeing peace and world cooperation in space: (1) Explicit confirmation that the U.N. Charter applies to the limits of space exploration; (2) a declaration that space and heavenly bodies are not subject to claim of national sovereignty; (3) an international system for registering of all objects launched into space; (4) a specialized space unit in the United Nations Secretariat; (5) a world weather watch using satellites; (6) a cooperative search for ways toward weather modification; and (7) a global system of communications to link the world by telegraph, telephone, radio, and television.

October 22: National Science Foundation announced the establishment of a science resources planning office to study U.S. long-range scientific needs, to be headed by NSF Associate Director for Planning, Richard H. Bolt.

October 23: The Freedom 7 Mercury capsule in which Alan B. Shepard, Jr., made the first suborbital space flight, was presented to the National Air Museum of the Smithsonian Institution. In his presentation, NASA Administrator Webb said: "To Americans seeking answers, proof that man can survive in the hostile realm of space is not enough. A solid and meaningful foundation for public support and the basis for our Apollo man-in-space effort is that U.S. astronauts are going into space to do useful work in the cause of all their fellow men."

"Such flights as those of Freedom 7 are not stunts. They are not antithetical to sober scientific and technological research. Interpreted properly, these dramatic events can add much to public understanding and excite creative interest in extending the base on which public support must rest."

—— NASA announced that it had ordered 14 additional Delta launch vehicles (Douglas Thor first stage, Aerojet-General AJ10–118 second stage, and Allegheny Ballistics Laboratory third stage) for Relay, Syncom, Telstar, and Tiros satellites. Five of the first six of the twelve Deltas successfully launched Echo I, Tiros II and III, and Explorers X and XII.

—— Ranger launching again postponed at Atlantic Missile Range because of technical difficulties.

—— Cleveland extension (SNPO–C) of the joint AEC-NASA Space Nuclear Propulsion Office (SNPO) activated, located on Lewis Research Center and headed by John L. Wilson.

—— USAF Discoverer XXXIII failed to achieve polar orbit.

—— First underwater launching of Navy Polaris A–2, and first firing from submarine, U.S.S. *Ethan Allen.*

—— AEC announced that the Soviet Union had detonated a thermonuclear bomb with a 30-megaton yield as well as a small underwater nuclear device. These were the 22d and 23d Soviet nuclear tests reported by AEC.

—— Marshal R. Y. Malinovsky, Soviet Defense Minister, announced that the U.S.S.R. had solved the problem of antimissile defense (a statement later qualified in retranslation).

October 24: Studies of "unconventional" rockets using liquid fuels in the thrust range from 2 to 24 million pounds announced by NASA; 2 contracts being carried out by Aerojet-General and Rocketdyne Division of North American Aviation.

—— USAF Titan fired from Cape Canaveral to coincide with overhead passage of Midas IV.

—— The first Centaur liquid-oxygen/liquid-nitrogen tanking tests were successfully completed at Sycamore Canyon.

—— Long duration static test of the S–I stage (SA–2 vehicle) occurred at Marshall Space Flight Center, for a period of 120 seconds.

—— Small liquid-fuel rocket was fueled and fired while floating in ocean off Point Mugu in Aerojet-General demonstration of this launching technique.

October 25: NASA selected Pearl River site in southwestern Mississippi, 35 miles from the Michoud plant in New Orleans, for static test facility for Saturn and Nova-class vehicles, completed facility to operate under direction of Marshall Space Flight Center.

——— Ranger 2 shot postponed indefinitely as the 8-day "window" (i.e., when the Moon, Sun, and Earth were in favorable positions) had ended before technical difficulties could be corrected.

——— Reported from Cape Canaveral that launch of Titan the previous evening had been detected by Midas IV.

——— USAF announced that Project West Ford's 350 million dipoles launched with Midas IV had not yet been found by radar contact.

——— Full Tass text of Marshal R. V. Malinovsky's speech on October 23 as it appeared in Soviet dailies, showed no statement to the effect that the Soviets had perfected an antimissile missile, as had been reported by Moscow correspondents of the American press.

October 26: National Aeronautic Association appointed committee headed by Maj. Gen. Albert Boyd, U.S. Air Force (retired), to program U.S. efforts to break world aircraft records now held by other nations.

——— Air Force Cambridge Research Laboratory successfully flight-tested largest plastic balloon (318 feet in diameter and 434 feet long).

——— U.S.S. *Blandy* demonstrated capability of a destroyer to recover MR-2 Mercury capsule, with Virgil Grissom aboard, from water in series of pickups in lower Chesapeake Bay.

October 27: Largest known rocket launch to date, the Saturn 1st stage booster, successful on first test flight from Atlantic Missile Range. With its eight clustered engines developing almost 1.3 million pounds of thrust at launch, the Saturn (SA-1) hurled water-filled dummy upper stages to an altitude of 84.8 miles and 214.7 miles down range. In a postlaunch statement, Administrator Webb said: "The flight today was a splendid demonstration of the strength of our national space program and an important milestone in the buildup of our national capacity to launch heavy payloads necessary to carry out the program projected by President Kennedy on May 25. We in NASA deeply appreciate the contribution by the military services and American industry in achieving this important milestone. * * *" Development of Saturn had begun under Advanced Research Projects Agency auspices in 1958.

——— Goddard Space Flight Center and Geophysics Corp. launched Nike-Cajun rocket from Wallops Station with 60-pound payload that reached 90-mile altitude in a study of electron density and temperature in the upper level of the atmosphere.

——— Smithsonian Astrophysical Observatory scientists reported that Discoverer XVII nose cone (launched November 12, 1960, from Pacific Missile Range) had picked up tritium—product of a solar flare in which hydrogen and helium combine at high energies. While recovered satellite capsules often pick up some tritium, capsule of Discoverer XVII had 100 times the normal amount.

October 27: Secretary of Defense McNamara announced that progress of the Administration's accelerated defense buildup made unnecessary the use of additional defense funds appropriated by the Congress above the amount requested by the administration. The Congress had voted $514.5 million for additional long-range bombers; $180 million additional for the B-70; and $85.8 million additional for Dyna-Soar.

—— Second NASA honor awards ceremony in Washington: Dr. Abe Silverstein, new Director of the Lewis Research Center, received NASA's Outstanding Leadership Medal; William O'Sullivan of Project Echo received the NASA Exceptional Scientific Achievement Award; and George D. McCauley received the Sustained Superior Performance Award. Other NASA personnel who had received NASA or non-Federal awards during NASA's third year were also recognized.

—— Comdr. Alan B. Shepard, Jr., awarded the Theodore Roosevelt Distinguished Service Award in New York City.

—— All-out speed trial of X-15 postponed because of heavy cloud cover, a flight aimed at 4,100 miles per hour.

October 29: NASA announced that first Mercury-Scout launch to verify the readiness of the worldwide Mercury tracking network would take place at Atlantic Missile Range.

—— U.S.S.R. announced completion of its series of Pacific rocket tests with a successful shot of 7,500 miles. Since series began on September 13, Tass had announced a total of eight shots, emphasizing the accuracy of what was described as a "fundamentally new type of guidance system."

October 30: U.S.S.R. exploded 55- to 60-megaton nuclear device as per Khrushchev's promise to the 22d Communist Party Congress. White House release later in the day pointed out that this Soviet explosion would "produce more radioactive fallout than any previous explosion. The Soviet explosion was a political rather than a military act."

October 31: NASA has assembled an outstanding management team for its stepped-up assault on space, NASA Administrator Webb told the Aero Club of Washington. Citing Associate Administrator Seamans, Deputy Associate Administrator Dixon, and Director of Manned Space Flight Holmes as examples, Mr. Webb added: "These men, and many others associated with them, know the technical side of aeronautics and space and are all experienced in the management of large activities. Each has demonstrated a personal earning capacity far beyond what the Government is able to pay for their services. Each is thoroughly familiar with the opportunities and problems associated with our most important technical military weapon system development efforts. It is fortunate for this Nation that men with these high qualifications and such experience are willing to forego large earnings in industry and a more normal personal and family life to supply the leadership needed in our national space effort."

—— Launch of Mercury-Scout canceled at T minus 10 seconds at Atlantic Missile Range because of mechanical difficulties, while record speed flight of X-15 was again prevented by cloud cover restricting instrumentation.

October 31: At autumn meeting of the National Academy of Sciences in Los Angeles, Dr. Hilde Kallmann-Biji reported on Committee on Space Research report by an international studies group on data discovered by Soviet and American satellites as well as sounding rocket observations including those of Britain and West Germany. Findings indicated that 500 miles in space, temperatures may fluctuate 1,000° F., and that the Earth's upper atmosphere has distinct day and night variations in density and pressure.

During October: A series of some 50 supersonic flights to analyze the characteristics, intensity, and air and ground effects of supersonic booms began at Edwards Air Force Base, Calif., under joint sponsorship of the USAF, FAA, and NASA.

NOVEMBER 1961

November 1: New organization of NASA headquarters became effective, which established four major program offices (Manned Space Flight, Space Sciences, Advanced Research and Technology, and Applications), and provided center directors with direct line to the Office of the Associate Administrator.

——— Reported that the DOD–NASA Golovin Committee was near agreement on hybrid solid-and-liquid fuel rockets for Dyna-Soar. Golovin Committee had been meeting for 3 months to work out families of large rockets for overall national space program.

——— Mercury-Scout, testing global tracking network, was destroyed by range safety officer after lift-off.

——— Radiocarbon from nuclear tests had been useful in tracing movements of the atmosphere, reported Prof. Gordon S. Fergusson to the National Academy of Sciences. Studies since 1955 showed that it took 1 year for carbon 14 to move from one hemisphere to the other, once it gets into the lower atmosphere.

——— Scientists and engineers of Langley Research Center and the Space Task Group were guests of the Peninsula Chamber of Commerce honoring the 46 years of NACA and NASA on the Virginia Peninsula.

——— Ballute (balloon braking system) reentry test of 500-pound Cree vehicle, launched by three-stage Nike rocket, reaching an altitude of 28 miles and a speed of near 1,900 miles per hour, at Eglin Air Force Base, Fla.

——— USAF Hound Dog missile launched successfully from B–52 over Atlantic Missile Range and hit target area.

November 2: Lewis Research Center scientists, G. B. Brown and E. E. Callighan, reported at 1961 International Conference on High Magnetic Fields at MIT, that NASA was constructing a magnetic shield of superconducting alloys for future manned spacecraft.

——— British Skylark reached an altitude of more than 100 miles in third of four NASA firings to study ultraviolet radiation in the Southern Hemisphere.

——— Navy Aerobee 100, which had been launched from water on October 24, recovered and overhauled, made its second successful launch at Point Mugu, Calif. This was the second successful launch of liquid-fuel rocket from the open sea, having been towed to sea, fueled horizontally, ballasted to float vertically, and ignited by remote control.

——— Reported unnamed NASA spokesman stated that two inspections of Gurtler-Hebert & Co.'s renovation of the Michoud Ordnance Plant for NASA revealed no racial discrimination.

November 3: NASA announced start of a nationwide recruiting drive for 2,000 talented scientists and engineers.

November 3: Nine-nation Western European Conference in London announced decision to launch a satellite in mid-1965, using a British Blue Streak first stage, a French Veronique second stage, and a West German third stage, from the Woomera range in Australia.

——— Lincoln Laboratory of MIT reported that examination of telemetry from Midas IV indicated that Project West Ford package of dipoles had been ejected at the expected time and the proper speed. No evidence was available as to whether the dipoles had been released, and no radar reading had been obtained.

——— Three Polaris A-2 missiles successfully fired within 3-hour period from submarine *Ethan Allen.*

November 4: USAF Office of Aerospace Research symposium at MIT, at which Dr. Otto Schmitt, of Northwestern University, reported that snails, worms, and one-celled paramecia had the ability to detect magnetic fields encountered on the surface of the Earth.

November 5: USAF Discoverer XXXIV launched into polar orbit with recoverable capsule. Launch represented 22d successful in the Discoverer series.

November 6: NASA informed Marshall Space Flight Center that management of the Agena B vehicle system would be retained at Marshall Space Flight Center.

——— Department of Commerce issued a proposal by MIT researchers on a science information network, entitled "An Experimental Communications Center for Scientific and Technical Information" (OTS, AD-255626). Proposed network included newspapers as well as radio and TV and recommended further research on a specific network to process, identify and retrieve scientific documents and information for dissemination.

——— N. Varvarov, in Soviet newspaper, Ekonomicheskaya Gazeta, denounced the U.S. space program as using outer space for military purposes and cluttering the cosmos with an unnecessarily large number of satellites. Especially critical of the Discoverer series, the article said: "* * * the United States, pursuing an intensive arms race, is setting up an elaborate system of cosmic military intelligence communications and navigation. * * * Actually, this is banditry on an international scale. * * *"

November 7: Explosion in hydrogen system canceled full-power run of AEC Kiwi B-1A reactor at Jackass Flats, Nev. Five men were injured, and the reactor was not damaged.

——— NASA announced award of a contract to North American Aviation Co. to study the feasibility of a large erectable manned space station based on Langley Research Center concept.

November 8: Industry proposal to FCC for organizing a commercial communications satellite system critically reviewed in hearings of the Monopoly Subcommittee of the Senate Small Business Committee.

November 9: X-15 flown to announced record 4,070 miles per hour (later revised to 4,093) by Maj. Robert White, U.S. Air Force, in top-speed test flight, making safe landing with outer right windshield cracked.

November 9: Former President Dwight D. Eisenhower in speech at Case Institute of Technology in Cleveland, said:
"I for one do not fully understand why, in the midst of a plethora of necessary and costly activities, our Nation should be required, urgently, to develop a capacity to put men on the Moon and challenge our principal opponent in doing so. * * *"
—— Senate Small Business Committee concluded critical hearings on the FCC's handling of the communications satellite project.
November 10: USAF Atlas with capsule containing squirrel monkey destroyed by range safety officer at Atlantic Missile Range when main sustainer engine failed 15 seconds after launch.
—— Reported that ONR-supported radio observatory at Cal Tech's Owens Valley, Calif., had expanded the radius of the observable universe 27 times (36 sextillion miles, the distance traveled by light at a speed of about 186,000 miles a second in 6 billion years).
—— In reviewing NASA's communications satellite programs, Administrator Webb pointed out that it had been speculated that the satellite system "may have progressed enough by 1964 that we shall be able to watch the Tokyo Olympic Games on television at home."
November 11: NASA announced that top speed of X-15 on Major White's record flight was revised to 4,093 miles per hour (mach 6.04), reached at 95,800 feet. (White also held altitude record of 217,000 feet (41 miles), flown on October 11).
November 12: Mercury-Atlas 5, scheduled for launch no earlier than November 14, ran into technical difficulties, postponing launch for several days.
—— Bell Aerosystems Co. announced design of a "practical zero gravity belt" to propel a man a short distance in space.
November 13: USAF announced that amateur radio communications satellite, assembled by Project Oscar Association, would be flown piggyback in future Discoverer vehicle.
November 13-22: International Meteorological Satellite Workshop held in Washington, D.C., attended by weathermen from 28 nations, sponsored by NASA and the U.S. Weather Bureau to apply Tiros-acquired data to practical day-to-day weather prognosis.
November 14: United Arab Republic neither confirmed nor denied reports of November 8 that it had successfully launched its first rocket. Dr. Eugen Saenger of the Stuttgart Jet Propulsion Institute in Germany denied any connection with the United Arab Republic program as charged by Israel.
—— Soviet and bloc delegates from Czechoslovakia and Poland, who had previously accepted invitations, did not attend the NASA-Weather Bureau International Meteorological Satellite Workshop held in Washington; telegram from Andre Zolotukhin received by Dr. Reichelderfer stated that "our representatives unable to participate," but requested "dispatch of relevant papers if possible."
November 15: Navy Transit IV-B and Traac (Transit Research Attitude Control) satellites launched into orbit by Thor-Able-Star at Atlantic Missile Range.
—— USAF Discoverer XXXV launched into polar orbit with 300-pound recoverable capsule.

November 15: NASA Bios (biological investigation of space) payload launched by Argo D-8 booster rocket from Pacific Missile Range, but veered sharply off course 57 seconds after launch.

—— Anniversary of the first flight of the USAF-USN-NASA X-15 powered with the XLR-99 engine (15 flights total to date). A $225 million research program under NASA management, test data indicated that X-15 would exceed its design limits by 100 percent in altitude and 17 percent in speed. The X-15 had already pushed near its design altitude limit of 250,000 feet (reached 217,000 feet October 11, 1961) and passed its maximum design speed of mach 6 (reached mach 6.04 November 9, 1961).

—— NASA Director of the Office of Manned Space Flight, D. Brainerd Holmes, said, in an interview, that at least 10 Apollo spacecraft would be ordered in the manned lunar vehicle prime systems contract to be awarded in December 1961.

—— Army launched Speedball rocket successfully from the island of Roi-Namur on Kwajalein atoll in the Southwest Pacific, the first target rocket to be used in Nike-Zeus development.

November 16: Gold-plated capsule of Discoverer XXXV recovered after 18 orbits in midair over Fern Island by C-130 aircraft, Capt. James F. McCullough, U.S. Air Force, as pilot. It was the 10th recovery from orbit in the Discoverer series and the 1st recovery observed from the ground.

—— In speech on "Scientists and Engineers in the Space Program," Albert F. Siepert, NASA Director of Administration, outlined NASA's basic policies on personnel. He pointed out that of NASA's some 20,000 employees, only 4,000 had come to NASA through individual appointments, the remainder on transfer of organizations intact to NASA. NASA's personnel utilization practices, Siepert said, were as follows: (1) Don't use a scientist or engineer when another skill will do as well; (2) classify a man's skills by what he actually does, rather than how he was formally trained; (3) provide professional entrance into the Federal civil service through an examination which is work centered rather than academically oriented; (4) take on-the-job training and education seriously; (5) encourage professional recognition outside the agency; and (6) recognize that job satisfaction depends upon the man's continued interest in his work as well as his take-home pay."

—— William J. O'Sullivan, Jr., of Langley Research Center awarded the Second NASA Invention and Contribution Award for conception and development of the inflatable space vehicle. Proposed in January 1956 to the U.S. IGY Committee, O'Sullivan's invention led to two successful NASA experiments, Echo I and Explorer IX, and U.S. Patent No. 2,996,212, entitled "Self-Supporting Space Vehicle" issued to the NASA Administrator in behalf of the United States on August 15, 1961.

—— Army Nike-Zeus antimissile rocket with active second stage successfully fired at Point Mugu, Calif.

November 17: NASA announced selection of the Chrysler Corp. for construction, test, and launch of 20 first-stage Saturn boosters at its Michoud, La., fabrication plant.

—— First USAF Minuteman successfully fired from silo at Atlantic Missile Range, making 3,000-mile flight.

November 18: Ranger II placed into low orbit from Atlantic Missile Range by Atlas, but Agena second stage did not restart, leaving deep-space probe Ranger in parking orbit. Results reported to delay lunar-landing Ranger shot in early 1962.

—— NASA announced that record Argo D-8 vehicle was launched with Bios payload from Point Arguello, but reentry capsule beacon signal had not been acquired by down-range recovery forces.

—— Evidence of traces of living things in meteorites from space reaching Earth, reported in Nature magazine by George Claus of NYU and Bartholomew Nagy of Fordham, based upon discovery of five types of "organized structure" in the Orgueil meteorite found in southern France in 1864, and the Ivuna meteorite that fell in central Africa in 1938.

—— Reported from Moscow that U.S.S.R. was planning to orbit a man around the Moon in 1962, and that the U.S.S.R. had ICBM's in being with 100-megaton warheads.

November 19: NASA announced the completion of the preliminary flight rating test of the Nation's first liquid-hydrogen rocket engine. The engine, the RL-10, was designed and developed by Pratt and Whitney, of United Aircraft, for the Marshall Space Flight Center, and 20 captive firings were completed within 5 days under simulated space conditions, consistently producing 15,000 pounds of thrust. RL-10, previously known as XLR-115, was initiated in October 1958 and over 700 firings were conducted in its development.

—— Navy Skylark balloon began coast-to-coast flight carrying University of Chicago cosmic ray experiment, launched at Brawley, Calif., and landing near Asheville, N.C., on November 21.

November 20: NASA announced consolidation of nuclear-electric propulsion program at Lewis Research Center by transfer of the Marshall Space Flight Center Research Projects Division under Dr. Ernst Stuhlinger to Lewis within 3 months.

—— Executive order of the President suspended the 8-hour limitation on construction workers in NASA. It stated that "a clearly leading role in aeronautical and space achievement has become a vital national objective," and that it was essential to conduct the space program "with a major national commitment of manpower, material, and facilities," and "with all possible speed and efficiency."

—— In news conference, Dr. Albert R. Hibbs, Jet Propulsion Laboratory's Chief of Space Sciences, stated that it would be a "major accomplishment" if the United States were to overtake the Russians in the race to the Moon, a "less than 50-50 chance." He pointed to the rumors that the Soviet Union had already attempted to launch a probe to Mars.

—— NASA Launch Operations Directorate announced establishment of Offices of Financial Management and of Procurement and Contracts to support NASA activities at Cape Canaveral, previously done by Marshall Space Flight Center.

November 20-21: Technical conference on the progress of X-15 research held at Edwards Air Force Base, sponsored jointly by NASA, USAF, and USN; the third in a series, previously held in 1956 and 1958.

November 21: In a speech on "Our National Program in Space," NASA Administrator Webb said:

"In carrying out its responsibilities, NASA cooperates with and depends upon private industry, universities, and many other Government agencies—not only the Department of Defense, the Atomic Energy Commission, and the Bureau of Standards, but the Weather Bureau, the Federal Communications Commission, the Federal Aviation Agency, the National Science Foundation, and others.

"It has been only 4 years since the first manmade satellite orbited the Earth. Since then, progress in this new field of space has been tremendous. I believe that in the years ahead the rate of progress will trace a steeply ascending curve. I believe also that the many problems we will solve to achieve manned exploration of space will create a wealth of new materials, consumer goods, processes, and techniques, thus opening a host of new jobs, careers, opportunities for investment, and a general national growth.

"We can be first in space if we advance our scientific and technical knowledge at the most rapid rate possible, and if we go forward with the sustained effort that it requires. That is the basis of our national space effort."

—— Titan ICBM launched from Cape Canaveral carrying target nose cone to be used in Nike-Zeus antimissile-missile tests. This was first Titan ICBM to be fired from Cape Canaveral by a military crew, AFBSD's 6555th Aerospace Test Wing.

November 22: USAF launched an unnamed satellite with an Atlas-Agena booster from Point Arguello, Calif., in first unannounced U.S. satellite launching.

—— An F4H Phantom II piloted by Lt. Col. Robert B. Robinson (USMC), claimed a new world speed record at Edwards Air Force Base, averaging 1,606.324 miles per hour.

November 23: Tiros II completed first year in orbit, still transmitting cloud-cover photographs of usable quality, although it has been expected to have a useful lifetime of only 3 months. Tiros II had completed 5,354 orbits, and had transmitted over 36,000 photographs.

—— National Aeronautic Association notified Mrs. Constance Wolf, of Blue Bell, Pa., that her Texas-to-Oklahoma balloon flight of 40 hours 13 minutes, 363.99 miles and 13,000-foot altitude established 15 women's world records.

November 24: First four U.S. Nike-Cajun rockets arrived in Norway for use in research program off Andoeya Island early next year.

—— DOD announced that the Advanced Research Projects Agency (ARPA) had selected Space Technology Laboratory (STL) to develop satellite system under Project Vela for detection of clandestine nuclear weapons in outer space.

—— Official Soviet films on the flight of Vostok II shown on nationwide TV in NBC program, "Crossing the Threshold—Part I."

November 25: Announced that the largest quartz lens ever ground had been completed by Bausch & Lomb for use in NASA's optical solar simulation system at the Jet Propulsion Laboratory. The lens is 36 inches in diameter, 6 inches thick at the center, and weighs 350 pounds.

November 26: Russian scientist, K. Florensky, reported in Komsomol Pravda that a comet's head, not a meteorite or an interplanetary atomic weapon, caused the big explosion that jarred Siberia on June 30, 1908. The blast near the Hunguska River killed 1,500 reindeer, felled trees over an area of 700 square miles, and was recorded on seismographs around the world.

November 27: The United States presented an outline for a program for cooperation and control in outer space to the U.N. Committee on the Peaceful Uses of Outer Space meeting in New York. This was the first committee meeting held since its membership was expanded from 18 to 24 nations by the General Assembly 2 years ago, the 7 Communist members refusing to attend until today. U.N. Delegate Charles W. Yost urged consideration of the U.S. proposals before the life of this Committee expired at the end of the year. The U.S. proposals were: acknowledge that international law and U.N. Charter extend to outer space; establish central registry for all space launchings and satellites; and share all information from weather satellites.

—— USAF reported that Lincoln Laboratory's Millstone Hill radar at West Ford, Mass., had made three sightings on November 3, 13, and 15, which might be the missing package of 350 million copper needles launched into orbit on October 21.

—— Senator Robert Kerr announced that he would introduce legislation to authorize private ownership of the U.S. portion of the proposed worldwide communications satellite system. His bill would create the "Satellite Communications Corp." which the participating firms would buy.

—— The Daniel and Florence Guggenheim Foundation announced openings for 18 young scientists and engineers for graduate study in rockets, jet propulsion, space flight and space structures at Cal Tech's Jet Propulsion Laboratory, Aerospace Laboratory at Princeton, and the Institute of Flight Structures at Columbia. Established in 1949, Guggenheim fellowship program in the aerospace sciences has provided financial aid to 142 students to date.

—— General Curtis E. LeMay, Chief of Staff, U.S. Air Force, said in interview with U.S. News & World Report:

"I think we're at the period of space technology that we were in aeronautical technology along about 1914. At that time no one could predict the type of weapon systems that the airplane was going to produce, or the transportation system that it would eventually produce. As a matter of fact, it was pretty much a laughing stock—a very expensive toy.

"We, of course, initially did a very poor job in our development program to advance the science of aeronautics in this country. I hope we do a better job in space. At the present time we can't predict what will eventually come out of research in space in the way of weapon systems or commercial vehicles or any other use that we might put space to. I am sure that men are going out into space. I'm sure that they'll find useful things to do out there, and I'm sure that, unless something is done to preclude it, they'll find things to fight over out there, too."

November 28: President Kennedy awarded the Harmon International Aviator's Trophy jointly to the three X-15 test pilots. The first joint award in the history of the trophy went to A. Scott Crossfield of North American, Joseph A. Walker of NASA, and Maj. Robert M. White, U.S. Air Force.

—— NASA selected North American Aviation to design and build a three-man Apollo spacecraft leading toward eventual lunar landings and exploration of the Moon. Each proposal was evaluated by a team of nearly 200 NASA and DOD specialists.

November 29: Mercury-Atlas 5 launch from Cape Canaveral placed Mercury spacecraft carrying chimpanzee "Enos" into orbit; retro-rockets were fired on second rather than planned third orbit because of developing malfunction of attitude control system. Mercury capsule was recovered 1 hour and 25 minutes after water landing by the destroyer *Stormes*, and well-performing "Enos" recovered in excellent condition. Project Mercury officials named John H. Glenn as prime astronaut for the first manned orbital mission with M. Scott Carpenter as backup, and Donald Slayton as prime astronaut for second manned orbital mission with Walter Schirra as backup.

—— President Kennedy, after giving lengthy answer to a question at his regular press conference, was handed a note by his press secretary, which he read and then said:

"This chimpanzee who is flying in space took off at 10:08. He reports that everything is perfect and working well."

—— Thomas F. Dixon, NASA Deputy Associate Administrator, in speech before the Greater Los Angeles Press Club, reviewed west coast space projects and said:

"All of these projects are part of a unified national program, which was accelerated earlier this year. I want to emphasize that this is a national program. It is not just a NASA program. It is not just a government program. It is a program to mobilize America's manpower and resources to meet the goals we have set for ourselves in space."

—— Soviet Cosmonaut Gagarin in New Delhi said that "we will not have to wait long" for the first manned flight to the Moon. Gagarin was making a 9-day visit to India.

—— Air Force Office of Aerospace Research (OAR) announced that its Office of Scientific Research had awarded 139 basic research grants and contracts worth almost $8 million so far this fiscal year.

November 30: Army fired its Pershing solid-fuel tactical missile from Cape Canaveral on a 200-mile flight, testing accuracy, warhead components, and blast and heat factors at launch in relation to operational crew protection. This was the seventh straight successful firing of the Pershing.

—— Army successfully fired a Nike-Zeus antimissile missile from White Sands Missile Range in the first flight test of all three rocket motors.

During November: Studies by General Electric's Space Sciences Laboratory, under NASA contract, disclosed that the heat barrier encountered by vehicles returning from deep space will be at least 2½ times more severe than previously estimated.

During November: Textron's Bell Aerospace Corp. completed 81 flight tests with cold gas one-man propulsion system in USAF C-131 aircraft flying "Keplerian trajectories."

—— National Bureau of Standards established the Radio Refractive Index Data Center at its Boulder, Colo., laboratories, to correlate data from 300 reporting points on the variable refraction of radio waves at specific times, heights, and locations.

—— DOD revised its patent policy on space research and development contracts in accordance with present NASA patent provisions, such provisions already having been written into space communications contracts (i.e., Government retains royalty-free exclusive title to patents developed under contract).

—— USAF announced expansion of gaseous physics research activities with the construction of a $636,000 laboratory at L. G. Hanscom Field, Bedford, Mass., as a part of the Cambridge Research Laboratory.

—— Project Rover, Project Pluto, and the U.S. underground nuclear test program were halted in Nevada by a jurisdictional strike between the Operating Engineers and the Plumbers and Pipefitters Unions.

—— Representatives of 30 American aerospace firms in Europe formed an informal organization known as U.S. Aerospace Industries in Europe.

—— Douglas Aircraft reported successful drop and recovery of a data capsule and camera that will be used to film inflation of Echo-type spheres as a part of Project Big Shot (the first phase in the NASA program leading to a global communication system using rigidized inflatable spheres equidistant and in orbit around the Earth).

—— Two-hundred-foot radiotelescope of the Commonwealth Scientific and Industrial Research Organization was commissioned at Parkes, 200 miles west of Sydney, Australia. Slightly smaller than the British radiotelescope at Jodrell Bank, the Parkes telescope is considered superior in surface accuracy and tracking control. It cost $1.8 million of which the Rockefeller Foundation and the Carnegie Corp. donated $500,000.

—— USAF aircraft produced sonic booms on routine training missions over major airlane intersections, in support of FAA studies of supersonic air transportation problems.

DECEMBER 1961

December 1: Three new world helicopter speed records were claimed by Capt. Bruce K. Lloyd, U.S. Navy, and Comdr. E. J. Roulstone, U.S. Navy, who flew an HSS-2 helicopter at 182.8, 179.5, and 175.3 miles per hour for 100, 500, and 1,000 kilometers, respectively, over a course along Long Island Sound between Milford and Westbrook, Conn.

—— Nike-Zeus guidance system successfully passed initial flight test at White Sands Missile Range.

—— Navy-sponsored Hypersonic Propulsion Research Laboratory, for simulating flights at speeds up to mach 10, was opened at Applied Physics Laboratory of Johns Hopkins University.

December 1-2: Two Roksonde meteorological sounding rockets were successfully fired from Cape Canaveral, telemetered measurements of winds and temperatures at altitudes above 180,000 feet. Produced by Marquardt for the Army, Roksondes had already completed a series of tests at White Sands Missile Range and Pacific Missile Range.

December 2: Twelve Canadian Black Brant rockets for upper-atmosphere research were to be launched from NASA's Wallops Station, Virginia, as the Canadian Defence Research Board shifted the firing site from Fort Churchill because a fire largely destroyed the Canadian facilities. Capable of carrying a 150-pound payload to an altitude of 150 miles, Black Brants were to be fired from Wallops at the rate of two in December 1961, two in February 1962, six in April 1962, and two in May 1962.

December 4: Ambassador Adlai Stevenson introduced a resolution before the U.N.'s Political Committee for a U.N. space program guided by four considerations: (1) Application of the principles of international law to outer space and celestial bodies to ensure against sovereignty claims in space; (2) making the U.N. a clearinghouse for use of outer space, including information on satellite launchings and cooperation for peaceful use of outer space; (3) international cooperation on weather satellite information; (4) international cooperation on communications satellites.

Ambassador Stevenson said: "There is a right way and a wrong way to get on with the business of space exploration. In our judgment, the wrong way is to allow the march of science to become a runaway race into the unknown.

"The right way is to make it an ordered, peaceful and cooperative and constructive forward march under the aegis of the United Nations."

—— Reported from Cape Canaveral that Astronaut John H. Glenn, Jr., had moved into "ready room" quarters. NASA had made no announcement whether a man would ride in the next Mercury capsule.

—— USAF fired a Blue Scout rocket from Point Arguello, Calif., aimed at a point some 27,600 miles out in space and over the South Pole, to measure low-energy protons originating from the Sun.

December 5: A new world aircraft altitude record for sustained horizontal flight was claimed by Comdr. George W. Ellis, U.S. Navy, who flew an F4H Phantom II at 66,443.8 feet over Edwards Air Force Base, Calif.

——— AEC–NASA Space Nuclear Propulsion Office (SNPO) selected the proposal of the Aetron Division of Aerojet-General Corp. as the basis for negotiating an architect and engineering contract for an $8 million downward-firing test stand for the Nerva engine. The Nerva would be used in nuclear rockets with a reactor derived from the Kiwi B test series.

——— Reported by Drew Pearson that CIA had warned that Russia "is preparing to launch a man around the moon in 60 days."

December 6: The first Project Mercury manned orbital flight, MA–6, was scheduled by NASA for early in 1962 after analysis of the data from the MA–5 chimpanzee orbital flight indicated that the Mercury-Atlas system and the tracking network were ready for manned orbital flight.

——— Astronauts Alan B. Shepard, Jr., commander, U.S. Navy, and Virgil I. Grissom, captain, U.S. Air Force, were awarded the first astronaut wings (almost identical design of a shooting star imposed on the traditional pilot's badge) in a joint ceremony by their respective services.

——— U.S.S.R. raised its expenditure on science by 12 percent in its 1962 budget. The Minister of Finance, Vasily Garbuzov, announced that the 1962 expenditure on science would be 4,300 million rubles ($4,773 million). Also announced was a 44-percent increase in the defense budget to 13,400 million rubles ($14,874 million).

——— Italian Air Force crew fired Jupiter IRBM from Atlantic Missile Range, the third such launching.

December 7: NASA postponed its projected manned orbital flight from December 1961 until early in 1962 because of minor problems with the cooling system and positioning devices in the Mercury capsule, Dr. Hugh Dryden, Deputy Administrator of NASA, said in a Baltimore interview. "You like to have a man go with everything just as near perfect as possible. This business is risky. You can't avoid this, but you can take all the precautions you know about."

——— Plans for the development of a two-man Mercury capsule were announced by Robert Gilruth, Director of NASA's Manned Spacecraft Center. The two-man capsule, to be built by McDonnel Aircraft Corp., would be similar in shape to the Mercury capsule but slightly larger and two to three times heavier. Its booster rocket was announced to be the USAF Titan II, scheduled for flight test early in 1962. One of the major objectives in the two-man capsule program would be a test of orbital rendezvous, in which the two-man capsule would be put into orbit by the Titan II and would attempt to rendezvous with an Agena stage put into orbit by an Atlas rocket. Total cost for a dozen two-man capsules plus boosters and other equipment was estimated at $500,000,000. Program name later announced as Gemini.

December 7: Power run completed the test series on the Kiwi B-1A reactor system being conducted at the Nevada Test Site by AEC's Los Alamos Scientific Laboratory. Fourth in a series of test reactors in the joint AEC-NASA nuclear rocket propulsion program, Kiwi B-1A was disassembled for examination at the conclusion of the test runs.

—— Second Atlas ICBM launched by SAC crew, from Vandenberg Air Force Base.

—— Preproposal conference on the contract for design, research, development, fabrication, and testing of the reactor-in-flight-test (Rift) vehicle was held at Marshall Space Flight Center. This vehicle would test-fly the Nerva nuclear engine now under development. Twenty-nine firms were invited to attend this preliminary conference at which they were furnished general information on the project. Interested firms would then have 30 days to file information on their capabilities and experience. Then a smaller number of firms would be invited to submit detailed bids. Purpose of the two-step evaluation was to enable firms not in a competitive position to avoid the expense of entering detailed proposals.

—— United States and Soviet delegates to the United Nations informally discussed the question of the political makeup of the U.N. Committee on the Peaceful Uses of Outer Space and on a possible joint resolution in that Committee.

December 8: NASA selected Mason-Rust as the contractor to provide support services at NASA's Michoud plant near New Orleans, providing housekeeping services through June 30, 1962, for the three contractors who would produce the Saturn S-I and S-IB boosters and the Rift nuclear upper-stage vehicle.

—— USAF fired an Aerobee sounding rocket from Point Arguello, Calif., out over the Pacific 1,300 miles high and 900 miles toward Hawaii, at which point the rocket released three sets of flares to be photographed from California, Hawaii, and Alaska. Purpose was to provide a more precise knowledge of the location of Hawaii with respect to the North American mainland by means of photogrammetric triangulation of the flare photos.

December 9: Solid-propellent rocket motor generating nearly 500,000 pounds of thrust was fired in a static test of 80-second duration by United Technology Corp. at Sunnyvale, Calif., under USAF contract.

—— Nike-Zeus antimissile missile was fired from Point Mugu in its first low-altitude flight, going up to 40,000 feet and then out over the Pacific Missile Range at that altitude.

December 10: The Carnegie Institution issued annual report containing several findings from its space scientists: Philip Abelson contended that it was a waste of time and money to sterilize vehicles going to the Moon or planets because any life there would be so unlike terrestrial life that it could not be contaminated by Earth organisms; Horace Babcock offered a theory on alternating spiral magnetic fields of the Sun that might explain sunspots, flares, and the 22-year magnetic cycle; other scientists noted growing evidence of major differences in chemical composition of distant stars, indicating a need to revise methods of computing distances to those stars.

December 11: The U.N.'s Political Committee unanimously approved a resolution calling on the Committee on Peaceful Uses of Outer Space to meet on March 31, 1962, to begin discussions of world cooperation in space. The resolution essentially incorporated the four-point U.S. program on the peaceful uses of outer space. The U.S.S.R. supported the resolution although it had previously rendered the Committee inoperative by boycotting its meetings.

—— The national space program portends a major technological advance for mankind, NASA Associate Administrator Dr. Robert C. Seamans, Jr., told the New Orleans Chamber of Commerce. Comparing its potential to that of the invention of the steam engine, Dr. Seamans noted:

"Two aspects of such major advances are characteristic. First, the practical results are largely unforeseeable, primarily because they develop on broad fronts and, frequently, in unsuspected directions. Second, the concentration of effort required does not diminish effort expended on other frontiers of knowledge, but rather spurs such activities. For example, despite fears that space technology would monopolize the scientific effort of this country, such fields of activity as oceanography, geophysics, and the physics of high-energy particles have greatly increased since the national space effort has become a serious one."

—— Contract awarded by Army Engineers to Brown & Root, Inc., for design of major portion of NASA's Manned Spacecraft Center at Houston, Tex.

—— Survey of leading space experts on U.S. space goals from 1970–75 by the North American Newspaper Alliance produced a consensus that the United States would establish a Moon base from which to thoroughly explore the Moon and to launch interplanetary manned probes. Those interviewed included important figures in space industry, USAF, NASA, and space research.

December 12: Discoverer XXXVI was launched by the USAF into orbit from Vandenberg Air Force Base, Calif., with a piggyback 10-pound Oscar (orbiting satellite carrying amateur radio) satellite aboard in addition to the Discoverer payload. Oscar was the first satellite built by private citizens to be put in orbit, transmitted Morse signal to world amateur radio operators.

—— The National Center for Atmospheric Research was inaugurated at Boulder, Colo. To include the facilities of the High Altitude Observatory of the University of Colorado, to be governed by the University Corporation for Atmospheric Research, a corporation of 14 universities from coast to coast, and to be financed by the National Science Foundation, the center would provide a national attack on weather research, including the use of tools such as rockets, balloons, and computers too expensive for any one university to finance.

—— USAF Atlas launched from Atlantic Missile Range carried piggyback package of 28 dummy fuel cells in a study of how metals evaporate on reentry.

—— Army announced that track radar for Nike-Zeus antimissile missile had successfully tracked an Atlas ICBM on November 22 from Ascension Island as well as Echo I 1,500 miles from Earth.

December 13: NASA Administrator James E. Webb said in a speech in Cleveland that the United States would follow its first manned orbital flight in January 1962 with similar manned orbital flights every 60 days. These would gather data on effects of weightlessness, needed to determine the pacing of the two-man flight program later on. Mr. Webb also forecast the launching of 200 sounding rockets, 20 scientific satellites, and 2 deep-space probes in 1962.

—— USAF completed Titan I research and development test flight program of 40 launches at the Atlantic Missile Range; of the 40 launches 4 had been failures.

December 14: NASA fired a four-stage solid-fuel Trailblazer rocket from Wallops Station, Virginia, in the first of a series of reentry tests. Two stages boosted the rocket to 167 miles; then the other two drove the nose cone down through the atmosphere at 14,000 miles per hour.

—— Nike-Zeus firing in extended range from Point Mugu attained all test objectives.

December 15: NASA's Explorer XII satellite returned voluminous data revising previous information on the Van Allen radiation belts and showing them to be no substantial problem to manned space flight. Launched on August 15, 1961, and transmitting until December 6, 1961, Explorer XII returned information amounting to 5,636 telemetry tapes (2,400 feet each). Of principal interest was its finding that the Van Allen belts consist of a preponderance of protons over electrons in a ratio of 1,000 to 1. Since the protons are of less than 1 million electron volts energy, they do not themselves offer a serious radiation problem and serve to slow the velocity of other radiation.

—— S-IB stage of the Advanced Saturn launch vehicle would be built by the Boeing Co., NASA announced. The $300 million contract, to run through 1966, called for development, construction, and test of 24 flight stages, plus several for ground tests. Assembly would take place at the NASA Michoud Operations Plant, New Orleans, La. The S-IB would be the first stage of the vehicle that would launch the three-man Apollo spacecraft for direct circumlunar flight or, with rendezvous, for lunar landing.

—— In a ceremony at Edwards Air Force Base, Calif., the USAF graduated its first class of five pilot-engineers from its school for space pilots. Graduates were awarded advanced technical degrees.

December 18: Dr. Robert Jastrow, Chief of Theoretical Division and Director of the Institute for Space Studies, Goddard Space Flight Center, making the 25th annual Wright Brothers' Lecture before the Institute of Aerospace Sciences, reviewed progress in the space sciences, said the most exciting and fruitful area thus far had been investigation of "solar control over the atmosphere of the Earth, causes of weather activity in the lower atmosphere, and the structure of the upper atmosphere."

—— NASA announced that the first station in a network of data-gathering stations for use with second-generation satellites had been completed near Fairbanks, Alaska. Site for the second of the $5 million installations, each with a high-gain antenna 85 feet in diameter, was announced to be Rosman, N.C., 40 miles southwest of Asheville.

December 18: USAF awarded an additional $52 million contract to North American Aviation for development of a prototype B-70 bomber, bringing to $267 million the amount allocated for the B-70.

——— Capsule from Discoverer XXXVI was ejected from orbit after 6 days and a record of 64 orbits, landed in the Pacific near Hawaii, was kept afloat by 3 USAF pararescue men until arrival of Navy destroyer.

——— USAF Minuteman ICBM successfully fired from a silo and traveled 3,600 miles down the Atlantic Missile Range, the second consecutive successful silo launching.

——— Successful test of a new way to steer large-size rockets was announced by United Technology Corp., an experimental method called liquid thrust vector control (TVC), in which a gas or liquid is sprayed into the exhaust path of a rocket engine, deflecting the exhaust and thereby turning the vehicle. The test was made with a 450,000-pound-thrust solid-fuel engine.

——— DOD summary statement on the X-15 program stated that to that date there had been 45 flights of the X-15, with planned performance achieved on 42 and the prime research objectives achieved on 40. The 98-percent launch success record of the X-15 was attributed to (1) use of alternate modes for subsystems and (2) the presence of a pilot to detect malfunctions in subsystems. This compared to a 43-percent launch record for an unmanned missile with no alternate modes in subsystems.

December 19: NASA announced that Ira H. Abbott, Director of Advanced Research and Technology, would retire in January after 32 years service with NACA and NASA. Beginning with the Langley Aeronautical Laboratory in 1929, Abbott became internationally known for his aerodynamic research, in more recent years as supervisor of X-15, supersonic transport, nuclear rocket, and advanced reentry development programs.

December 19-20: A technical conference on problems of runway slush in winter jet operations was held in Washington under joint sponsorship of the Federal Aviation Agency and NASA. The conference, open to aviation representatives, reviewed the extensive research flight tests conducted at FAA's National Aviation Facilities Experimental Center and other experimental and theoretical work done at NASA's Langley Research Center. Interest in all experiments centered on the adverse effects of runway slush on takeoff and landing characteristics of jet aircraft. Research findings were that on both takeoff and landing in heavy slush jetliners tend to act like "a sailboat without a keel," that at takeoff speeds heavy slush causes jetliners to lose the effect of nose wheel steering and most of their braking power. Recommendations included the devising of a quick and accurate means of measuring runway slush and suspension of jet operations when slush reached a depth of 1 inch.

December 20: X-15 No. 3 made first flight, a successful test of new automated control system, NASA's Neil A. Armstrong as pilot in his first flight of XLR-99-engined X-15. At half throttle, X-15 reached speed of 2,502 miles per hour and an altitude of 81,000 feet.

December 20: NASA announced that Douglas Aircraft had been selected for negotiation of a contract to modify the Saturn S-IV stage by installing a single 200,000-pound-thrust, Rocketdyne J-2 liquid-hydrogen/liquid-oxygen engine instead of six 15,000-pound-thrust P. & W. hydrogen/oxygen engines. Known as S-IVB, this modified stage will be used in advanced Saturn configurations for manned circumlunar Apollo missions.

——— Two new radiotelescopes, one at Cambridge University and the other at Jodrell Bank, would be constructed with grants from Britain's Department of Scientific and Industrial Research totaling $3,360,000. The Cambridge telescope would consist of three 52-foot paraboloidal aerials, two fixed and one rail-mounted, designed to examine a limited area of the sky with greater precision than present equipment. The Jodrell addition would be a 125-foot telescope to be used in conjunction with the present 250-foot telescope.

——— USAF launched Atlas ICBM from Cape Canaveral with a rhesus monkey in a side-mounted pod on a flight 5,000 miles long and 600 miles in altitude. The flight was intended to produce information on reactions to launch and reentry conditions much more severe than in human flights. The monkey survived the flight but recovery attempts failed.

——— In San Bernardino news conference, Gen. Bernard Schriever, U.S. Air Force, said: "I have never felt we were behind Russia in missile development."

December 21: Army Nike-Zeus antimissile missile successfully intercepted a Nike-Hercules missile flying at over 3,000 miles per hour over White Sands Missile Range, while another Nike-Zeus made highest flight to date from Point Mugu and another Nike-Zeus was launched from Kwajalein Island in the South Pacific.

December 22: Unnamed USAF satellite launched from Point Arguello, Calif. The announcement said it was powered by an Atlas-Agena B combination and that the satellite was "carrying a number of classified test components."

——— NASA selected Air Products & Chemicals to supply additional liquid hydrogen for west coast development projects, a $35 million contract to be negotiated for a 5-year period.

December 26: Development time schedule for Dyna-Soar was reduced when DOD authorized the USAF to move directly from B-52 drop tests to unmanned and then manned orbital flights. This eliminated the previous interim stage of suborbital flights to be powered by the Titan II booster. USAF announced it was canceling the Titan II development contract held by the Martin Co. and negotiating a new contract for a larger booster.

——— Ten scientific organizations recommended that the American Association for the Advancement of Science create a new section to deal with scientific information and communication. The problem was seen to be one of an overabundance of information not accessible for the scientist, particularly in interdisciplinary science. Only three other new sections have been created in the American Association for the Advancement of Science in this century.

December 27: The "race in space" between the United States and U.S.S.R. was the top news story of 1961, with the Berlin crisis running second, according to a poll of Associated Press member newspapers and radio stations.

—— Dr. Glenn T. Seaborg, Chairman of the Atomic Energy Commission, speaking before the American Association for the Advancement of Science, said that although science had become a determining factor in national and international events, its effectiveness was hampered because educated men did not understand science. Yet, he said, science was truly a part of the humanities; "Who in our times can make an adequate criticism of life without knowledge of the ideals, the methods, the dynamics of science?"

—— DOD and USAF revealed that the B-70 bomber may be redesignated RSB-70 (reconnaissance-strike-bomber) and its mission changed from tracking known, fixed targets to seeking out and destroying unknown, hidden, or uncertain targets.

—— Dr. Carl Sagen, of University of California (Berkeley), disputed the "space seed" life theory in American Association for the Advancement of Science paper. "Panspermic" theory did not seem plausible in the light of the fierce environment of space and the vastness of the universe, he said.

December 28: Titan II, an advanced ICBM and the booster designated for NASA's two-man orbital flights, was successfully captive-fired for the first time at the Martin Co.'s Denver facilities. The test not only tested the flight vehicle but the checkout and launch equipment intended for operational use.

December 29: Dr. Hugh L. Dryden, Deputy Administrator of NASA, speaking in Denver before the American Association for the Advancement of Science, said: "The sheer magnitude of the manned lunar exploration program, amounting as it will to $3 billion or more [in fiscal year 1963], represents a significant application of the Nation's resources. These billions of dollars will be spent in the laboratories, workshops, and factories of the Nation and thus constitute a significant factor in the Nation's employment and economy generally. The personnel in the space program are not all scientists and engineers but come from every walk of life.

"The ultimate and practical purpose of these large expenditures is twofold: (1) Insurance of the Nation against scientific and technological obsolescence in a time of explosive advances in science and technology; and (2) insurance against the hazard of military surprise in space."

—— Dr. Joseph F. Shea was appointed Deputy Director for Systems Engineering, Office of Manned Space Flight, NASA headquarters, reporting to D. Brainerd Holmes, NASA's Director of Manned Space Flight. Dr. Shea came to NASA from Space Technology Laboratories.

—— Dr. Arthur Rudolph was appointed Assistant Director of Systems Engineering in NASA's Office of Manned Space Flight. Operating out of the Marshall Space Flight Center, Dr. Rudolph would serve as liaison between vehicle development at Marshall and the manned space flight program at the Manned Spacecraft Center in Houston.

December 30: Navy HSS-2 Sea King helicopter flown at 199 miles per hour for 3-kilometer distance claimed world record at Windsor Locks, Conn., by Commander P. L. Sullivan, U.S. Navy, and Capt. D. A. Spurlock, U.S. Marine Corps.

December 31: NASA established a Management Council to ensure the orderly and timely progress in the manned space flight programs. The Council, composed of senior officials from NASA headquarters, Marshall Space Flight Center, and the Manned Spacecraft Center, and chaired by D. Brainerd Holmes, Director of the Office of Manned Space Flight, would meet at least once a month to identify and resolve problems as early as possible and to coordinate the interface problems.

—— Dr. Robert C. Seamans, Jr., NASA Associate Administrator, said in a radio interview that a second Venus probe had been added to NASA's 1962 program as insurance for the first probe scheduled in August. Both probes would be the Mariner R, the reduced-weight version resorted to because of time slippage in the Centaur booster program. Dr. Seamans also said the United States plans three attempts to land instrumented packages on the Moon in 1962.

During December: General Electric announced operation of the largest solar thermionic power system at GE's solar test facility near Phoenix, Ariz. Early tests generated an output of 12.18 watts and unit has potential efficiency of 15 to 20 percent of the total solar energy input.

—— West German Post Office indicated that it would construct near Munich a ground station capable of handling up to 600 phone calls simultaneously for operations in late 1963 or early 1964 with Telstar and Relay type satellites.

—— Japan's launch facilities for its rocket research program would be moved from Akita on the northwest coast of Honshu to Kagoshima on the southern tip of Kiushu, according to an announcement by Hideo Itokawa at the Thul International Symposium on Rockets and Astronautics in Tokyo.

—— USIA reported that U.S. space achievements were a leading item in their overseas information program and covered all media. USIA concluded: "The policy of 'openness' observed in both U.S. manned space flights during the year dramatized the basic difference between the American open society and the Soviet closed society, and drew widespread approval from commentators throughout the free world. The availability of full information about the events through all news media, together with the presence of foreign correspondents—who gave firsthand, on-the-spot coverage—enabled oversea audiences to achieve a high degree of self-identification with one of the greatest adventures of our times."

APPENDIX A

SATELLITES, SPACE PROBES, AND MANNED SPACE FLIGHTS—A CHRONICLE FOR 1961

The following chart was prepared from open public sources by Dr. Frank W. Anderson, Jr., Deputy NASA Historian. Sources included "Satellite Situation Reports" prepared by the Space Operations Control Center at Goddard Space Flight Center and the "Space Activity Summaries" prepared by NASA's Office of Public Information. Russian data are unofficial.

It is a characteristic of documenting space flights that the data processing of telemetry requires time, sometimes in the order of months, until scientific results are available. The documentation of such results thus remains a problem resolutely attacked but not fully solved. Comments are invited.

Launch date	Name	International designation	Vehicle	Payload data	Apogee (statute miles)	Perigee (statute miles)	Period (minutes)	Inclination	Remarks
1961 Jan. 31	Samos II (United States).	1961 Alpha 1.	Atlas-Agena.	Total weight: 4,100 pounds, including 2d-stage casing. Objective: Determine capabilities for making observations of space, the atmosphere, and the nature of the globe from satellites. Payload: Photographic and related test equipment.	340	295	94.9	97.40	Near-circular polar orbit achieved.
Feb. 4	Sputnik IV (U.S.S.R.).	1961 Beta 1.	Not disclosed.	Total weight: 14,292 pounds. Objective: Develop and place heavy space vehicle in precise orbit. Payload: Not disclosed.	203.56	138.75	89.8	64° to Equator.	Decayed on Feb. 26, 1961.
12	Sputnik V (U.S.S.R.).	1961 Gamma 3.	----do----	Total weight: Not disclosed. Objective: Test injection of probe into interplanetary orbit. Payload: Venus probe.	198	123	89.7	Not disclosed.	Decayed on Feb. 25, 1961.
	Venus probe (U.S.S.R.).	1961 Gamma 1.	Sputnik V.	Total weight: 1,419 pounds. Objective: Probe in vicinity of Venus; test long-range communications; provide measurements and observations of solar system. Payload: Equipment to measure cosmic rays, magnetic fields, charged particles of interplanetary gas and corpuscular sunbeams, and to record micrometeor impacts.	1.0190 astronomical units.	.7183 astronomical unit.	300 days	0.581	Venus probe launched from orbiting Sputnik V; last radio contact Feb. 12, 1961, attempts at contact abandoned on June 15, 1961. Was predicted by U.S.S.R. to pass within 62,500 miles of Venus on May 19–20, 1961.
16	Explorer IX (United States).	1961 Delta 1.	Scout.	Total weight: 80 pounds, including 15-pound sphere and 65 pounds of ejection, inflation, telemetry, and other equipment. Objective: Study performance of Scout research rocket; orbit inflatable sphere to measure atmospheric density. Payload: 3.65-meter inflatable sphere of milar and aluminum foil; radio beacon, 280 solar cells, and miniature batteries.	1,605	395	118.3	38.63°	1st satellite orbited by all-solid-fuel rocket; first satellite orbited from Wallops Station. Radio beacon on balloon satellite never operated, requiring optical tracking. Early results: at 700 kilometers, atmospheric density was measured as 3×10^{-17} gram per cubic centimeter.
17	Discoverer XX (United States).	1961 Epsilon 1	Thor-Agena B.	Total weight: 2,450 pounds including 2d-stage casing and 300-pound reentry capsule, retrorocket, and recovery aids. Objective: Evaluation of Agena B satellite and of modified satellite stabilization system. Payload: As above, plus 4 silicon samples to test effects of high-energy protons on silicon used on transistors and solar cells.	486	177	95.4	80.43° to Equator.	Near-polar orbit achieved; capsule could not be recovered because of equipment malfunction in satellite. Still in orbit, Jan. 2, 1962.

AERONAUTICAL AND ASTRONAUTICAL EVENTS OF 1961

Date	Name	Launch vehicle	Payload and objective	Weight (pounds)	Apogee-Perigee	Period (min)	Angle to Equator	Remarks
18	Discoverer XXI (United States).	do	Total weight: 2,200 pounds (approximately). Objective: Collect data on atmospheric phenomena and infrared radiation in support of USAF early warning satellites.	650	149	93.8; 97.8 after restart.	80.74° to Equator.	Near-polar orbit achieved; 1st successful restart of Agena engine in space was made while satellite was in orbit. Satellite still in orbit.
21	Transit III-B and Lofti (United States).	Thor-Able-Star.	Payload: Not disclosed. Total weight: 250 pounds. Objective: Test of all-weather global navigational satellite system for ships and aircraft. Payload: 36-inch sphere ringed with 6,600 solar cells and containing 2 transmitters, command system, memory system, telemetry system, despin system, SECOR experiment. Total weight: 67 pounds.	511	117	94.5	28.36° to Equator.	Orbit achieved was elliptical instead of near circular as intended and hampered quality of Transit data; the 2 satellites failed to separate as planned but both transmitters functioned. Decayed on Mar. 30, 1961.
	Lofti		Objective: Measure intensity of very low frequency signals coming through the ionosphere. Payload: 20-inch sphere containing telemetry and transmitting equipment and 2 very low frequency receivers. Total weight: 10,340 pounds.					
Mar. 9	Spacecraft IV (U.S.S.R.).	Not disclosed.	Objective: Testing of structure and systems for manned spaceship, including exposure of animals to cosmic radiation. Payload: Telemetry and television equipment; reportedly carried 1 dog, guinea pigs, black mice, insects, and plant seeds.	155	115	Not disclosed.	Not disclosed.	Orbit achieved, capsule said recovered in predetermined area in U.S.S.R. on Mar. 9, 1961; animal passengers reported alive and well.
25	Explorer X (United States).	Thor-Delta.	Total weight: 79 pounds. Objective: Gather definitive data on interplanetary and Earth's magnetic fields and their interplay with solar plasma. Payload: 13-inch sphere connected by tube to 19-inch cylinder; magnetometer, 2 fluxgate magnetometers, plasma probe, optical aspect sensor, transmitter, and chemical batteries.	145,000	100	112 hours	32° to Equator.	Orbit achieved, transmitters functioned continuously for 60 hours; data supported theory that interplanetary magnetic field near Earth is mainly extension of Sun's magnetic field.
25	Spacecraft V (U.S.S.R.).	Not disclosed.	Total weight: 10,330 pounds. Objective: Further test of structure and systems for future manned spaceship, including capsule recovery with animals aboard. Payload: Radio tracking and communications equipment; TV to report on condition of animals; animal passengers, including at least 1 dog.	150	111	Calculated at 88.4.	64°54' to Equator.	Orbit achieved, cabin reported recovered in predetermined area in U.S.S.R., with animal passengers alive and well.

Launch date	Name	International designation	Vehicle	Payload data	Apogee (statute miles)	Perigee (statute miles)	Period (minutes)	Inclination	Remarks
1961 Apr. 8	Discoverer XXIII (United States).	1961 Lambda 1.	Thor-Agena B.	Total weight: 2,100 pounds, including 2d-stage casing and 300-pound reentry capsule, retrorocket, and recovery aids. Objective: Test of Agena B satellite, especially changes in guidance, stabilization, and propulsion for improved control of orbital period. Payload: Reentry capsule with retrorockets and recovery aids; external lights for tracking experiment.	882	126	101.2	81°94' to Equator.	Orbit achieved, capsule separated from Agena in orbit; capsule failed to return to Earth because of stabilization problems.
12	Vostok I (U.S.S.R.).	1961 Mu 1	Not disclosed.	Total weight: 10,418 pounds. Objective: Placing of manned spacecraft in orbit and recover man and spacecraft. Payload: In addition to man, life-support equipment, radio and television to relay information on condition of man.	187.66	108.76	89.1	65.07° to Equator.	Orbit achieved; spacecraft and pilot, Maj. Yuri Gagarin, recovered safely after 1 orbit of the Earth.
27	Explorer XI (United States).	1961 Nu 1	Juno II (4 stages).	Total weight: 82 pounds. Objective: Orbiting a gamma-ray telescope to detect high-energy gamma rays from cosmic sources and map their distribution. Payload: 12- by 23½-inch octagonal satellite mounted on 6- by 20½-inch instrument column; gamma-ray telescope, Sun and Earth sensors, micrometeorite shield, temperature sensor, damping mechanism; 2 radio transmitters, solar cells, and batteries.	1,113.2	304	108.1	28.8°	Orbit achieved, with all equipment functioning normally. Preliminary analysis of data from gamma-ray telescope measuring intensity of gamma radiation rules out 1 version of steady-state cosmology in which matter and antimatter were held to be created simultaneously. If this were correct the intensity of gamma radiation would have been 1,000 times greater than it measured.
May 5	Freedom 7 (MR-3) (United States).		Mercury-Redstone.	Total weight: 2,100 pounds (approximate), including man. Objective: Putting a man into suborbital flight and recovering man and spacecraft. Payload: In addition to man, life-support systems, radio, and telemetry equipment to report on condition of man.					Suborbital flight achieved; Astronaut Alan B. Shepard, Jr., went to 115 miles altitude, landed 297 miles down range, demonstrated pilot control during weightlessness; astronaut and capsule recovered by helicopter after 5 minutes in water and transferred to U.S.S. Champlain.
June 16	Discoverer XXV (United States).	1961 Xi 1	Thor-Agena B.	Total weight: 2,100 pounds (approximately), including 2d-stage casing and 300-pound reentry capsule with retrorockets and recovery aids. Objective: Testing of recently changed components in Agena B improving control of orbital period.	251.6	139.1	90.87	82.11° to Equator.	Orbit achieved, capsule ejected from orbit and recovered from sea that same day north of Hawaii.

AERONAUTICAL AND ASTRONAUTICAL EVENTS OF 1961

Date	Name	Designation	Vehicle	Payload/Objective	Weight	Period (min)	Inclination	Remarks
29	Transit IV-A (United States).	1961 Omicron 1.	Thor-Able-Star.	Payload: Rare and common metals for study of effects of space environment; radiation and micrometeorite measuring instruments. Total weight: No. 1, 1,175 pounds; No. 2, 56 pounds; No. 3, 40 pounds. Objective: Orbiting of 3 satellites, No. 1 to develop all-weather, global navigation system (Transit IV-A); No. 2 to measure solar X-ray radiation (Greb III); No. 3 to measure cosmic radiation and its intensity (Injun). Payload: No. 1, memory system and electronic clock, 4 transmitters; No. 2, 2 X-ray detectors, 1 transmitter; No. 3, 12 particle and proton detectors, 1 transmitter.	No. 1, 623; Nos. 2 and 3, 634.	No. 1, 534; Nos. 2 and 3, 534.	No. 1, 103.7; Nos. 2 and 3, 103.8.	67° to Equator.	Orbit achieved, but satellites 2 and 3 apparently failed to separate; this somewhat damaged the quality of transmission; Transit IV-A was 1st satellite to use nuclear generator. Transit IV-A data established that the Earth's equator is elliptical rather than round, there being a 1,000-foot difference between the longest and shortest equatorial diameter. This difference, if not allowed for, would affect satellite orbits.
July 7	Discoverer XXVI (United States).	1961 Pi 1	Thor-Agena B.	Total weight: 2,100 pounds (approximately) including 2d-stage casing and 300-pound reentry capsule with retrorockets and recovery aids. Objective: Evaluating of Agena B and new changes in components; improving of control of orbital period; ejection and recovery of capsule. Payload: Chemical elements, including silicon, iron, bismuth, yttrium, to be tested for effects of space environment; erosion gauge; micrometeorite detector.	503	146	95	82.93° to Equator.	Orbit achieved, capsule ejected some 3 hours later, recovered in midair northwest of Hawaii after 32 orbits.
12	Midas III (United States).	1961 Sigma 1	Atlas D-Agena B.	Total weight: 3,600 pounds (approximately) including entire 2d stage. Objective: Testing system for global detection of missile launchings. Payload: Sensor, telemetry, and communications equipment; details not released.	1,350	1,850	160	Not available.	Orbit achieved, almost perfectly circular; highest orbit to date, heaviest U.S. satellite to date.
12	Tiros III (United States).	1961 Rho 1	Thor-Delta.	Total weight: 285 pounds. Objective: Further development of satellite weather observation system; photographing Earth's cloud cover and transmitting it to Earth for analysis; measuring solar energy absorbed, reflected, and emitted by the Earth. Payload: 42- by 19-inch cylinder, containing 2 wide-angle cameras, 2 tape recorders and electronic clocks, 2 infrared cameras, 5 transmitters, attitude sensors.	506.44	461.02	100.4	47.8° to Equator.	Orbit achieved, cameras and infrared equipment transmitted good data.

84 AERONAUTICAL AND ASTRONAUTICAL EVENTS OF 1961

Launch date	Name	International designation	Vehicle	Payload data	Apogee (statute miles)	Perigee (statute miles)	Period (minutes)	Inclination	Remarks
1961 July 21	Liberty Bell 7 (MR-4) (United States)		Mercury-Redstone.	Total weight: 2,100 pounds. Objective: Further testing of Mercury capsule and life-support system in manned flight, preliminary to orbiting man. Payload: In addition to man, capsule, and life-support system, radio, and telemetry to report on condition of man.					Flight successful, reaching altitude of 118 miles and range of 303 miles. Pilot Virgil I. Grissom recovered, capsule partially flooded and lost. This was the 2d successful manned suborbital flight.
Aug. 6	Vostok II (U.S.S.R.)	1961 Tau 1	Not disclosed.	Total weight: 10,430 pounds. Objective: Orbit manned spacecraft for some 17 orbits and recover spacecraft and man; study effects of space environment on man and man's work performance during prolonged weightlessness.	115.3	110.3	88.6	64°5'36''	Orbit achieved; after 17½ orbits, spacecraft returned to Earth and both craft and pilot (Maj. Gherman Titov) successfully recovered.
15	Explorer XII (United States)	1961 Upsilon 1.	Thor-Delta.	Payload: In addition to man, included life-support equipment, radio, television, and telemetry to report on condition of man, tape recorder, automatic and manual control equipment. Total weight: 83 pounds. Objective: Investigating solar wind, interplanetary magnetic fields, distant parts of the Earth's magnetic field, energetic particles in interplanetary space and in the Van Allen radiation belts. Payload: 27- by 19-inch cone with magnetometer boom and 4 solar-cell paddles; 10 particle detection systems measuring proton and electron activity and its relationship to magnetic fields; optical attitude sensor; 1 transmitter.	47,800	180	26 hours, 25 minutes.	33.3°	Ceased transmitting Dec. 6, 1961, after 1,600,000,000 bits of data. Preliminary analysis of 10 percent of data indicated Van Allen belts really 1 magnetosphere—charged particles trapped in Earth's magnetic field; extends from 400 to 30,000 or 40,000 miles out, with abrupt outer boundary, topped by 20,000-kilometer electromagnetic turbulence. Magnetosphere's outer portion mostly low-energy protons, its electrons having less intensity than previously indicated; lower portion still dominated by high-energy protons.
23	Ranger I (United States).	1961 Phi 1	Atlas-Agena B.	Total weight: 675 pounds. Objective: Developing systems for lunar and interplanetary exploration; measuring cosmic rays, radiation, dust; checking on whether Earth is	312.5	105.3	91.1	32.9°	Orbit achieved was low, oval Earth orbit rather than deep space probe as planned; test of spacecraft achieved, but many of the desired interplanetary

AERONAUTICAL AND ASTRONAUTICAL EVENTS OF 1961   85

Date	Name	Designation	Vehicle	Payload/Objective	Weight	Perigee	Apogee	Inclination	Remarks
25	Explorer XIII (United States).	1961 Chi 1	Scout	followed by cometlike trail of hydrogen gas. Payload: 5- by 11-foot cone containing equipment for measuring solar radiation, energetic particles, cosmic rays, magnetic fields, solar X-rays, neutral hydrogen geocorona, cosmic dust, and friction; 2 transmitters, 8,680 solar cells, 1 silver zinc battery. Total weight: 187 pounds, including spent 50-pound 4th stage and 12-pound transition section. Objective: Testing performance of Scout vehicle and guidance; investigating the nature and effects of space flight on micrometeoroids.		Originally computed at 606.34; later at 722.	Originally computed at 174.60; later at 74.	Originally computed at 97.27; later at 97.5.	Originally computed at 36.42°, later at 37.5°.	measurements could not be made. Reentered Aug. 29, 1961. Orbit achieved was lower than planned; reentered on Aug. 27, 1961; all flight data recomputed to account for premature reentry (down from 1 year expected lifetime).
30	Discoverer XXIX (United States).	1961 Psi 1	Thor-Agena B.	Payload: 76- by 24-inch cylinder, almost covered by 5 types of micrometeoroid impact detectors; 2 transmitters, solar cells, and nickel cadmium batteries. Total weight: 2,100 pounds (approximate), including 2d-stage casing and 300-pound reentry capsule with retrorockets and recovery aids. Objective: Testing reliability of Agena B; improving control of orbital period; ejecting and recovering capsule. Payload: Test instruments to check on adjustments made as result of previous flights.		345	140	91	82.14°	Orbit achieved, capsule ejected on 33d orbit and recovered from sea north of Hawaii Sept. 4, 1961.
Sept. 12	Discoverer XXX (United States).	1961 Omega 1.	....do....	Total weight: 2,100 pounds (approximate), including 2d-stage casing and 300-pound reentry capsule with retrorockets and recovery aids. Objective: Testing reliability of Agena B; improving control of orbit; ejection and precision recovery of capsule. Payload: Biological samples, solar cells, and radiation-sensitive materials for testing of effects of exposure to radiation.		345	154	92.4	82.58°	Orbit achieved, capsule ejected on 33d orbit and recovered in midair near Hawaii Sept. 14, 1961.
13	Mercury-Atlas IV (United States).	1961 Alpha-Alpha 1.	Mercury-Atlas D (MA-4).	Total weight: 2,700 pounds in orbit (3,900 at liftoff, 2,200 at recovery) Objective: Orbiting the unmanned Mercury capsule to test systems and ability to return capsule to predetermined recovery area after 1 orbit; testing of global Mercury tracking network. Payload: Simulator of pilot, to test environmental control; 2 voice tapes to check tracking network; life-support system; 3 cameras, tape recorder, telemetry.		158.6	100	88.6	32.57°	Orbit achieved, reentry made automatically on program; capsule recovered within programed area; 161 miles east of Bermuda; all test objectives met.

Launch date	Name	International designation	Vehicle	Payload data	Apogee (statute miles)	Perigee (statute miles)	Period (minutes)	Inclination	Remarks
1961 Sept. 17	Discoverer XXXI (United States).	1961 Alpha-Beta 1.	Thor-Agena B.	Total weight: 2,100 pounds (approximate), including 2d-stage casing and 300-pound reentry capsule with retrorockets and recovery aids. Objective: Testing reliability of Agena B; improving control of orbital period; ejecting capsule for precision recovery. Payload: Test instruments to check on adjustments made as a result of previous flights.	255	152	91	82.7°	Orbit achieved, but 2d stage and capsule failed to separate; capsule did not reenter as programed; decayed Oct. 26, 1961.
Oct. 13	Discoverer XXXII (United States).	1961 Alpha-Gamma 1.	...do...	Total weight: 2,100 pounds (approximate), including 2d-stage casing and 300-pound reentry capsule with retrorockets and recovery aids. Objective: Testing reliability of Agena B; improving control of orbital period; ejection and recovery of capsule; investigating radiation effects. Payload: Test instruments to check on adjustments made as a result of previous flights; equipment for measuring radiation; samples for radiation testing, including metals, seed corn, shielding materials, and solar cells; transmitter to investigate effects of ionosphere on radio propagation.	246.06	147.07	90.84	81.69°	Orbit achieved, capsule ejected on 18tb orbit and recovered in midair north of Hawaii Oct. 14, 1961.
19	P-21 Probe (United States).		Scout	Total weight: 94 pounds. Objective: Testing Scout vehicle and guidance system, measuring ionospheric electron density profile, and testing new Doppler velocity, and position facility at Wallops Station. Payload: 15- by 33-inch 8-sided frustrum containing electron sensor, 2 transmitters, nickel cadmium batteries.	4,261				Attained planned altitude and transmitted good data. Confirmed Explorer VIII data on helium layer, provided a structure of ionosphere: At 120 kilometers a mixture of nitrogen and oxygen molecules; from 120 to 1,000 kilometers, predominately atomic oxygen; from 1,000 to 2,500 kilometers, helium; above 2,500 kilometers, hydrogen.

AERONAUTICAL AND ASTRONAUTICAL EVENTS OF 1961

21	Midas IV (United States).	1961 Alpha-Delta 1.	Atlas-Agena.	Total weight: 3,500 pounds (approximately), including 2d stage. Objective: Place experimental satellite in orbit and eject the piggyback West Ford dipoles. Payload: In addition to Midas equipment, 75-pound package of 350,000,000 copper dipoles to form radio-reflection belt.	Not available.	Not available.	172	Not available.	Orbit achieved; Project West Ford piggyback capsule ejected in orbit but fate uncertain; no definite radar sighting.
Nov. 5	Discoverer XXXIV (United States).	1961 Alpha-Epsilon 1.	Thor-Agena B.	Total weight: 2,100 pounds (approximately), including 2d-stage casing and 300-pound reentry capsule with retrorockets and recovery aids. Objective: Testing reliability of Agena B; improving control of orbital period; ejection and recovery of capsule. Payload: Unidentified instruments plus reentry capsule.	614	140	96.9	82.52°	Orbit achieved; capsule was ejected in orbit but failed to reenter because of in-orbit malfunction.
15	Transit IV-B and Traac (United States).	1961 Alpha-Eta 1.	Thor-Able-Star	Total weight: 200 pounds. Objective: Testing of all-weather global navigation satellite for use by ships, aircraft. Payload: 43- by 31-inch drum carrying stable oscillators, continuous transmitters, phase modulators, memory system, clock, Snap auxiliary power system for transmitters.	700	582	105.6	32.42°	Orbit achieved.
	Traac	1961 Alpha-Eta 2.		Total weight: 200 pounds. Objective: Testing a gravity orientation system for satellites; gaining data on Van Allen belts. Payload: 43- by 16-inch doorknob containing gravity gradient stabilization equipment, particle detectors.	720	562	105.6	32.43°	Do.
15	Discoverer XXXV (United States).	1961 Alpha-Zeta 1.	Thor-Agena B.	Total weight: 2,100 pounds (approximately), including 2d-stage casing and 300-pound reentry capsule with retrorockets and recovery aids. Objective: Testing design changes; obtain radiation data; obtain data on future spacecraft design; orbit, eject, and recover capsule. Payload: Instruments to measure effects of design changes; experiments on future design; radiation measuring equipment.	173.4	147.2	89.76	81.63°	Orbit achieved; capsule ejected and retrieved in midair by a C-130 some 500 miles northwest of Honolulu on Nov. 16, 1961.

Launch date	Name	International designation	Vehicle	Payload data	Apogee (statute miles)	Perigee (statute miles)	Period (minutes)	Inclination	Remarks
1961 Nov. 18	Ranger II (United States).	1961 Alpha-Theta 1.	Atlas-Agena.	Total weight: 675 pounds. Objective: Test spacecraft systems for future lunar and interplanetary missions; obtain data on cosmic rays, radiation, and dust particles and on whether the Earth is trailed by a cometlike tail of hydrogen gas. Payload: 5- by 11-foot cone with hexagonal base, containing experiments on solar radiation, particle detection, cosmic rays, magnetic fields, solar X-rays, neutral hydrogen geocorona, cosmic dust, and space friction; 2 transmitters, 8,680 solar cells.	145.7	94.9	88.3	33.3°	Orbit achieved was low Earth orbit rather than the deep elliptical one planned, so data was not obtained on some of the test items. Primary objective of testing the system was achieved. The satellite reentered on Nov. 18, 1961, somewhere between 6th and 13th orbit.
29	Mercury-Atlas V (MA-5) (United States).	1961 Alpha-Iota 1.	Atlas D	Total weight: 4,100 pounds, liftoff; 2,900 pounds, orbit; 2,400 pounds, recovery. Objective: Orbit Mercury spacecraft with chimpanzee aboard in test of all Mercury systems preparatory to manned orbital flight. Payload: In addition to chimpanzee, 4 cameras, 6 radiation measurement packs, 78 temperature measurement instruments, 2 playback tape recorders.	147.5	99.6	88.5	32.5°	Orbit achieved; on 2d of 3 planned orbits, Mercury capsule was programed into reentry because of minor difficulties with attitude jets and electrical systems; capsule landed in the recovery area near Puerto Rico about 1:28 p.m., Nov. 29, 1961, was picked up by U.S.S. *Stormes* about 2:53 p.m. the same day; chimpanzee Enos performed satisfactorily in flight and was in good condition on recovery. Flight was considered successful, no additional primate flights were felt necessary to prepare for the manned orbital flight.
Dec. 4	USAF Probe (United States).		Blue Scout Junior.	Total weight: 29 pounds. Objective: To measure low-energy protons coming from the Sun. Payload: 6 "open window" photomultiplier tube-type detectors; 2 solid-state detectors; new 12-pound telemetry system including quarter-watt transmitter, digilock encoder.	27,600				Flight successful. 1st flight of a Blue Scout rocket configuration from the Pacific Missile Range.

AERONAUTICAL AND ASTRONAUTICAL EVENTS OF 1961

12	Discoverer XXXVI (United States).	1961 Alpha-Kappa 1.	Thor-Agena B.	Total weight: 2,100 pounds (approximately) including 300-pound reentry capsule with retrorockets and recovery aids. Objective: To conduct radiation and radio experiments; to orbit Agena stage, Discoverer capsule, and Oscar satellite; to eject from orbit and recover the Discoverer capsule. Payload: Capsule contained radiation experiments including biopack of human and animal tissues, spores, molds, and algae; nuclear track plate; dosimeters and samples of silicon, bismuth, magnesium, gold, titanium, and nickel; radiation instruments including ones measuring the time history of radiation and radiation shielding properties of materials. Agena stage (to remain in orbit) contained radiation-measuring instruments relating to Advanced Research Projects Agency's Vela Hotel program, cosmic monitors, impedance probe, and radio propagation experiment. Also 10-pound Oscar (Orbiting Satellite Carrying Amateur Radio) experiment.	280	148	91.5	81.21°	Orbit achieved; Oscar was ejected into separate orbit and transmitted to worldwide "ham" radio network; Discoverer capsule was ejected from orbit Dec. 18, 1961, landed in the Pacific near Hawaii, was kept afloat by 3 USAF pararescue men until picked up by U.S.S. *Renshaw*.
	and								
	Oscar (United States).	1961 Alpha-Kappa 2.			258	146	91.1	81.21°	

APPENDIX B

X-15 FLIGHT LOG

Conceived in 1952 and begun as a NACA-USAF-USN program in 1954, the rocket research aircraft X-15 had achieved a highly productive record by the end of 1961. The following chronology of successful flights, prepared by NASA's Office of Public Information (62-95), provides a useful listing of flight research milestones with the X-15 through April 30, 1962.

X-15 flight log

Date	Flight No.[1]	Pilot	Mach No.	Maximum altitude (feet)	Remarks
1959					
June 8	1-1-5	Crossfield	0.79	37,550	Planned glide flight.
Sept. 17	2-1-3	do	2.11	52,341	1st powered flight.
Oct. 17	2-2-6	do	2.15	61,781	
Nov. 5	2-3-9	do	1.00	45,462	Engine fire; fuselage structural failure on landing.
1960					
Jan. 23	1-2-7	do	2.53	66,844	
Feb. 11	2-4-11	do	2.22	88,116	
Feb. 17	2-5-12	do	1.57	42,640	
Mar. 17	2-6-13	do	2.15	52,640	
Mar. 25	1-3-8	Walker	2.00	48,630	1st Government flight.
Mar. 29	2-7-15	Crossfield	1.96	49,982	
Mar. 31	2-8-16	do	2.03	51,356	
Apr. 13	1-4-9	White	1.94	48,000	
Apr. 19	1-5-10	Walker	2.56	59,496	
May 6	1-6-11	White	2.20	60,938	
May 12	1-7-12	Walker	3.19	77,882	
May 19	1-8-13	White	2.31	108,997	
May 26	2-9-18	Crossfield	2.20	51,282	
Aug. 4	1-9-17	Walker	3.31	78,112	2,196 m.p.h.
Aug. 12	1-10-19	White	2.52	136,500	
Aug. 19	1-11-21	Walker	3.13	75,982	1,986 m.p.h.
Sept. 10	1-12-23	White	3.23	79,864	2,182 m.p.h.
Sept. 23	1-13-25	Petersen	1.68	53,043	
Oct. 20	1-14-27	do	1.94	53,800	
Oct. 28	1-15-28	McKay	2.02	50,700	
Nov. 4	1-16-29	Rushworth	1.95	48,900	
Nov. 15	2-10-21	Crossfield	2.97	81,200	1st flight with XLR-99 design engine.
Nov. 17	1-17-30	Rushworth	1.90	54,750	
Nov. 22	2-11-22	Crossfield	2.51	61,900	1st restart with XLR-99 design engine.
Nov. 30	1-18-31	Armstrong	1.75	48,840	
Dec. 6	2-12-23	Crossfield	2.85	53,374	
Dec. 9	1-19-32	Armstrong	1.80	50,095	1st hot nose flight.
1961					
Feb. 1	1-20-35	McKay	1.88	49,780	
Feb. 7	1-21-36	White	3.50	78,150	2,275 m.p.h., last LR-11 flight.
Mar. 7	2-13-26	do	4.43	77,450	2,905 m.p.h.; 1st Government XLR-99 flight.
Mar. 30	2-14-28	Walker	3.95	169,500	2,760 m.p.h.
Apr. 21	2-15-29	White	4.62	105,000	3,074 m.p.h.
May 25	2-16-31	Walker	4.90	107,500	3,300 m.p.h.
June 23	2-17-33	White	5.27	107,700	3,603 m.p.h.
Aug. 10	1-22-37	Petersen	4.11	78,200	2,735 m.p.h.
Sept. 12	2-18-34	Walker	5.25	114,300	3,614 m.p.h.
Sept. 28	2-19-35	Petersen	5.30	100,800	3,600 m.p.h.
Oct. 4	1-23-39	Rushworth	4.30	78,000	2,830 m.p.h.; flight made with lower ventral off.
Oct. 11	2-20-36	White	5.21	217,000	3,647 m.p.h.; outer panel of left windshield cracked.
Oct. 17	1-24-40	Walker	5.74	108,600	3,900 m.p.h.
Nov. 9	2-21-37	White	6.04	101,600	4,093 m.p.h.; outer panel of right windshield cracked.
Dec. 20	3-1-2	Armstrong	3.76	81,000	2,502 m.p.h.; 1st flight for X-15 No. 3.
1962					
Jan. 10	1-25-44	Petersen	.97	44,750	645 m.p.h.; emergency landing on Mud Lake after engine failed to light.
Jan. 17	3-2-3	Armstrong	5.51	133,500	3,765 m.p.h.
Apr. 5	3-3-7	do	4.06	179,000	2,830 m.p.h.
Apr. 19	1-26-46	Walker	5.84	150,000	3,920 m.p.h.
Apr. 20	3-4-8	Armstrong	5.33	207,000	3,818 m.p.h.
Apr. 30	1-27-48	Walker	4.58	246,700	3,443 m.p.h.

[1] Flight activity code:
  1st number, X-15 airplane number.
  2d number, flight number for specified airplane.
  3d number, X-15/B-52 airborne mission number.

X-15 Pilots

A. Scott Crossfield, North American Aviation, Inc.
Joseph A. Walker, research pilot, NASA Flight Research Center.
Maj. Robert M. White, U.S. Air Force.
Comdr. Forrest S. Petersen, U.S. Navy.
John B. McKay, research pilot, NASA Flight Research Center.
Capt. Robert A. Rushworth, U.S. Air Force.
Neil A. Armstrong, research pilot, NASA Flight Research Center.

Box score

Aircraft	B-52/X-15 flights	X-15 launches
1	47	27
2	38	21
3	9	4
Total	94	[1] 52

[1] Includes 2 glide flights without power.

NOTE.—Project officials estimate mission primary objectives were attained on 49 of 52 X-15 flights.

INDEX

	Page
Abbott, Ira H	24, 48, 75
Abelson, Philip	72
Ablation	1
Able	10
Able-Star	10
Acoustic chamber	8
Ad Hoc Carrier Committee (FCC)	55
Ad Hoc Task Group (NASA)	26, 27
Advanced Research and Technology, Office of (NASA)	49, 61
Advanced Research Projects Agency (ARPA)	50, 58, 66
Aerobee	39
Hi	5, 11, 25
100	61
150	39
Aero Club	59
Aerojet-General Corp	24, 25, 39, 41, 43, 55, 57, 71
Aeronautical Research Laboratory (USAF)	21
Aeronautics and Astronautics Coordinating Board (AACB)	2
Aerospace Industries Association	23
Aerospace Laboratory (Princeton University)	67
Aerospace Medical Association	23
Aerospace Medicine, School of	47
Aetron	71
Africa	35, 65
Agena B	7, 9, 25, 32, 40, 45, 62
Air Force Association	49
Air Force Ballistic Systems Division (AFSC)	39, 66
Air Force Herald (U.S.S.R.)	55
Air Force Proving Ground Command (AFPGC)	20
Air Force Systems Command (AFSC)	14, 31, 47, 50
Air Materiel Command (USAF)	14
Airplane, invention of	49
Air Products and Chemicals	76
Air Training Command (USAF)	10, 47
Air travel	33
AJ-10-118	43, 57
Alabama	27
Alabama, University of	27, 52
Alaska	72
Albuquerque, N. Mex	10
Algae	13
Allegheny Ballistics Laboratory	57
Alouette	26
Alpha 1 (*Samos II*)	80
Alpha-Alpha 1 (*Mercury-Atlas IV*)	85
Alpha-Beta 1 (*Discoverer XXXI*)	86
Alpha-Delta 1 (*Midas IV*)	87
Alpha-Epsilon 1 (*Discoverer XXXIV*)	87
Alpha-Eta 1 (*Transit IV-B*)	87
Alpha-Eta 2 (*Traac*)	87
Alpha-Gamma 1 (*Discoverer XXXII*)	86
Alpha-Iota 1 (*Mercury-Atlas V*)	88
Alpha-Kappa 1 (*Discoverer XXXVI*)	89
Alpha-Kappa 2 (*Oscar*)	89
Alpha-particle	35
Alpha-Theta 1 (*Ranger II*)	88

INDEX

	Page
Alpha-Zeta 1 (*Discoverer XXXV*)	87
Altitude chamber	14
American Assembly	56
American Association for the Advancement of Science (AAAS)	76, 77
American Astronautical Society	23, 33
American Geophysical Union	16
American Institute of Biological Sciences	23, 66
American Institute of Industrial Engineering	52
American Physical Society	16
American Rocket Society	23, 53, 54
American Telephone & Telegraph Co. (A.T. & T.)	3, 15, 35
Ames Research Center	2, 5, 17, 43
Andoeya Island	66
"An Experimental Communications Center for Scientific and Technical Information"	62
Antarctica	45
Antenna	45
Antisubmarine	47
Antofagasta, Chile	42
Apollo	19, 20, 23, 33, 35, 57, 64, 68, 74, 76
Appel, Lt. B (USAF)	38
Applications Programs, Office of (NASA)	49, 61
Applied Physics Laboratory	70
Arcas	39
Arcas-Robin	20
Arctic Circle	39
Argo D-4	46, 54, 55
D-8	63, 65
Arizona State University	10
Armour Research Foundation	14
Armstrong, Neil A	75, 91, 92
Army Chemical Corps Biological Laboratories	23
Arnold Engineering Development Center (USAF)	4, 35
Ascension Island	73
Asheville, N.C.	31, 65, 74
Assistant Director for Nuclear Applications (NASA)	9
Associated Press	77
Astro-Electronics Division (RCA)	32
Astronaut	14, 19, 33, 63, 66
Astronautics Documentation Subcommittee (FAI)	34
Astronautics Foundation, Inc	30
Astronautics Medal	63
Astronaut wings	71
Astronomical unit	20, 29
Astronomical yardstick	15
Astrophysics	21
Atlantic Missile Range (AMR)	1, 3, 4, 5, 7, 9, 11, 16, 17, 19, 21, 22, 23, 24, 28, 30, 31, 32, 33, 34, 35, 37, 38, 40, 41, 42, 45, 46, 47, 50, 52, 53, 56, 57, 58, 59, 61, 63, 64, 65, 66, 68, 70, 71, 73, 74, 75, 76.
Atlantic Ocean	10, 24, 63
Atlas	63, 72, 73, 76
D	3, 10, 21, 23, 53, 65, 72
E	22, 31, 36, 61
F	38
Atlas-Agena	66, 80, 87, 88
Agena A	4
Agena B	41, 50, 65, 76, 83, 84
Atlas-Centaur	37
Atmospheric density	46
Atmospheric drag	39
Atmosperic pressure	32
Atomic Energy Commission (AEC)	5, 9, 12, 20, 31, 45, 47, 53, 57, 62, 66, 71, 72, 75
Attitude control	36
Auburn University	27
Avco	13

	Page
"Aviation and Cosmonautics" (U.S.S.R.)	55
"Aviatsiga I Kosmonavtika." (See "Aviation and Cosmonautics.")	
Awards	2, 20, 24, 26, 27, 33, 40, 50, 55, 59
Azusa, Calif	26
B-52	30, 61, 76
B-52H	1, 9
B-58 (Hustler)	2, 22, 24, 30
B-70	13, 59, 75, 77
Babcock, Horace	72
Baikonur, U.S.S.R	23
Ballistic missile early warning system (BMEWS)	48
Balloon	15, 17, 19, 21, 35, 45, 48, 58, 61, 65, 66, 73
Ballute	61
Baltimore, Md	56, 71
Bausch & Lomb	66
Beacon (satellite)	3
Bear. (See Tu-114.)	
Beckman Instruments	1
Bedford, Mass	69
Bell Aerospace Corp	69
Bell Aerosystems Co	25, 63
Bell, David E	11
Bell Telephone Laboratories	8, 22
Belmar, N.J	19
Bendix Trophy	22
Berkeley, Calif	40
Berkner, Dr. Lloyd V	23
Berlin	77
Bermuda	10
Beta I (*Sputnik IV*)	80
Bethesda Naval Hospital	47
Betsy, Hurricane	45
Big Shot, Project	69
Bikle, Paul F	8
Bioastronautics Division (AFSC)	47, 50
Biomedical	25
Bios	64, 65
Black Brant	70
Blaw Knox Co	41
Blimps	36
Blue Bell, Pa	66
Blue Scout	1, 9, 15, 40, 70
Blue Scout, Jr	88
Blue Streak	62
Boeing Co	1, 5, 13, 30, 31, 50, 74
Bolometer	21
Bolt, Richard H	57
Bomarc B	19, 47
Boston, Mass	30, 45
Boulder, Colo	69, 73
Boundary layer	24
Bowen, Dr. E. G	43
Boyd, Maj. Gen. Albert (USAF, retired)	58
Bradley Field, Conn	20
Brawley, Calif	65
Brazil	35
Bronk, Dr. Detlev	10, 48
Brooks Air Force Base, Tex	38, 50
Brooks, Representative Overton	47
Brown & Root, Inc	73
Brown, G. B	61
Brown, Dr. Harold	9, 56
Budget (NASA)	12, 22, 29, 33, 38, 40
Buffalo, University of	38
Bullpup	6
C-130B	46, 55, 64
C-131	69
Cabin pressure	29

INDEX

	Page
California	72
California Institute of Technology	63, 67
California, University of	9, 12, 40, 45
Callighan, E. E.	61
Cambridge Research Laboratory (USAF)	1, 15, 25, 29, 58, 69
Cambridge University	76
Canada	11, 26, 55, 70
Canadian Defence Research Board	70
Canton Island	14
Cape Canaveral (*see also* Atlantic Missile Range)	9, 21, 25, 29, 33, 36, 37, 39, 41, 44, 65
Cape Town, South Africa	31
Capitol	20
Carbon	16
Carla, Hurricane	45
Carnegie Corp	69
Carnegie Institution	72
Carpenter, Lt. Comdr. M. Scott (USN)	68
Carswell AFB, Texas	22
Case Institute of Technology	63
Castro, Fidel	34
"Celestial Simulator"	36
Centaur (*see also* Atlas-Centaur)	4, 5, 35, 40, 50, 52, 57, 78
Central Aero Club (U.S.S.R.)	34
Central Committee (U.S.S.R.)	38
Central Intelligence Agency (CIA)	71
Centrifuge	14
Chance Vought Corp	3, 5, 9
Chapman, Dr. Sydney	37
Chesapeake Bay	58
Chi I (*Explorer XIII*)	85
Chicago, Ill	21
Chicago, University of	35, 65
China (Communist)	28
China, Republic of	40
Chrysler Corp	5, 64
Claus, George	65
Cleveland Extension (SNPO-C)	57
Cleveland, Harlan	56
Cleveland, Ohio	63, 74
Cochran, Jacqueline	41, 63
Collier Trophy	20, 26
Colorado, University of	73
Columbia University	56, 67
Comet	67
Comision Nacionale de Investigaciones Espaciales (Argentina)	26
Commerce, Department of	62
"Commercial Applications of Space Communications Systems"	54
"Commercial Supersonic Transport Aircraft Report"	34
Committee for Space Research (France)	11
Committee on Atmospheric Sciences (NAS)	30
Commonwealth Scientific and Industrial Research Organization	69
"Communication Satellite Policy"	34
Communications	50
Communications satellite	1, 3, 4, 6, 7, 8, 14, 15, 21, 22, 27, 29, 32, 34, 35, 37, 39, 41, 46, 53, 54, 55, 62, 63, 67, 70, 73.
Compromise	37, 39
Congressional Medal	10
Constan, Dr. George N	50
Convair	2, 10
"Coordination of Information on Current Research and Development Supported by the U.S. Government"	15
Cornell University	16
Corps of Engineers	25, 48, 73
Cosmic ray	65
Cosmodrome	23

INDEX

	Page
Cosmic radiation	35
Cosmonaut	15, 17, 23, 34, 48, 49
Cospar	3, 14, 60
Council of Scientists (USAF)	51
Cox, Dr. Hiden T	56
Cracow Observatory	28
Cree	61
Crossfield, A. Scott	27, 68, 91, 92
Cyclotron	28
Czechoslovakia	63
Dahlgren, Va	5
Daniel and Florence Guggenheim Foundation	67
Daniel and Florence Guggenheim International Astronautics Award	40
DC-8	40
Dean, Arthur H	12
Debbie, Hurricane	45
Deep Space instrumentation facility (DSIF)	6, 41, 45
Deep space tracking station	1, 6, 45
DeEsch, Lt. Earl H	42
Defense, Department of	2, 7, 9, 11, 12, 14, 22, 28, 29, 32, 33, 34, 38, 41, 43, 47, 56, 61, 66, 68, 69, 76, 77.
Defense Science Board	38
Delta	10, 43, 57
Delta I *(Explorer IX)*	80
Denver, Colo	77
Department of Science and Industrial Research (United Kingdom)	11, 76
Deutschendorf, Maj. H. J	2
Dillon, C. Douglas	22
Direct ascent	4
Director of Research and Engineering (DOD)	9
Disarmament	12
Discoverer program	31, 41, 62, 63, 64
Discoverer XVII	58
Discoverer XX	7, 80
Discoverer XXI	7, 81
Discoverer XXII	12, 42
Discoverer XXIII	14, 15, 82
Discoverer XXIV	25
Discoverer XXV	27, 82
Discoverer XXVI	31, 83
Discoverer XXVII	33
Discoverer XXVIII	37
Discoverer XXIX	42, 44, 85
Discoverer XXX	46, 85
Discoverer XXXI	47, 86
Discoverer XXXII	55, 86
Discoverer XXXIII	57
Discoverer XXXIV	62, 87
Discoverer XXXV	63, 64, 87
Discoverer XXXVI	73, 75, 89
Disorientation	53
Distinguished Service Medal (NASA)	20, 33
Dixon, Thomas F	39, 49, 59, 68
DOD. (*See* Defense, Department of.)	
Doenhoff, Albert E	24
Dolan, Charles J	10
Douglas Aircraft Co	1, 2, 28, 31, 40, 43, 57, 69, 76
Douglas, Senator Paul H	33
Downing, Representative Thomas N	48
Downward-firing test stand	71
Draper, C. Stark	38
Drop test	76
Dryden, Dr. Hugh L	6, 20, 21, 24, 28, 52, 71, 77
"Dumping Profile"	45
Dynamic test stand	15
Dyna-Soar	3, 12, 18, 30, 33, 50, 59, 61, 76

INDEX

	Page
E-66A	18
E-166	53
Earth	6, 10, 15, 16, 20
Echo, Project	7, 59, 69
Echo I	4, 8, 15, 22, 25, 26, 39, 43, 57, 64, 73
Edwards Air Force Base, Calif	20, 22, 38, 41, 46, 54, 60, 65, 66, 71, 74
Edwards, Calif	5, 6, 15, 16, 50, 55
Edwards Rocket Test Center	32
Eglin Air Force Base, Fla	20, 29, 47, 61
Eglin Gulf Test Range	19
Ehricke, Krafft	55
Eisenhower, President Dwight D	1, 2, 63
Ekonomicheskaya Gazeta	62
Electric propulsion	65
Electron	35, 58, 74
Electronic Industries Association	23
El Campo, Tex	16
Electrons	17
Elliott Cresson Medal	56
Ellis, Comdr. George W. (USN)	71
Ellith, G	55
El Segundo, Calif	28
Engine maintenance and disassembly building	42
England. (*See* United Kingdom.)	
Enos	68, 88
Ent Air Force Base, Colo	6
Environmental Simulation Laboratory (Naval Missile Center)	8
Epsilon (*Discoverer XX*)	80
Esso Research	10
Esther, Hurricane	45
Eta (*Transit III-B*)	81
Europe	21, 69
Satellite launcher	4
Executive Committee for Joint Lunar Study (NASA-DOD)	28
Exceptional Scientific Achievement Award (NASA)	59
Experimental Text Pilots, Society of	50
Explorer VII	55
Explorer VIII	26
Explorer IX	7, 64, 80
Explorer X	11, 12, 16, 43, 57, 81
Explorer XI	17, 82
Explorer XII	39, 40, 43, 57, 74, 84
Explorer XIII	41, 42, 85
F-1 engine	6, 14, 32, 40
F4H (Phantom II)	22, 42, 66, 71
Fairbanks, Alaska	74
Falkland Islands	11
Federal Aviation Agency (FAA)	12, 34, 60, 66, 69, 75
Federal Communications Commission (FCC)	2, 3, 8, 21, 55, 62, 63, 66
Fédération Aeronautique Internationale. (*See* International Aeronautical Federation.)	
Feldman, George J	53
Fellows, Col. Scott (USAF)	28
Fergusson, Gordon S	61
Fern Island	64
Finger, Harold B	9
First National Conference on the Peaceful Uses of Space	23
Flin Flon, Manitoba	45
Florence, Italy	14
Florensky, K	67
Ford Foundation	45
Fordham University	10, 65
Fort Churchill, Canada	70
Fort Eustis, Va	25
France	4, 7, 31, 46, 62, 65
Commission for Spatial and Scientific Research	2, 11
Rocket launching	7

INDEX

	Page
Franklin Institute	56
Freedom 7 (MR-3)	19, 23, 26, 57, 82
Frictional drag	24
Frontiers of Science Foundation (Oklahoma)	23
Frutkin, Arnold	61
Fuel cell	73
Full-pressure suit	29
Furnas, Dr. Clifford C	38
G-forces	19
Gagarin, Maj. Yuri A	15, 19, 21, 23, 27, 31, 32, 34, 35, 42, 44, 48, 50
Gaithersburg, Md	26
Galactic noise	10
Gallup poll	24
Gamma 1	80
Gamma 3 (*Sputnik V*)	80
Gamma ray	17
GAM-83B	6
Garber, Paul	21
Garbuzov, Vasily	71
Gazenko, O. G	53
G. Edward Pendray Award	55
Gemini, Project	71
General Dynamics/Astronautics Corp	33
General Electric Co	1, 6, 8, 13, 20, 33, 68, 78
Geneva, Switzerland	12, 23
Geodesy	14
Geophysics	73
Geophysics Corp	58
Geophysics Research Board (NAS)	11, 30
George Washington University	25
Gilpatric, Roswell	7
Gilruth, Robert R	46, 48, 71
Glenn, Lt. Col. John H., Jr. (USMC)	7, 68, 70
Glennan, T. Keith	2, 5
Glider	8
Goddard, Dr. Robert H	10, 55
Goddard Institute for Space Studies (GISS)	3, 10, 21, 74
Goddard, Mrs. Robert H	10
Goddard Space Flight Center	3, 10, 12, 17, 25, 32, 41, 58, 74
Goldberg, Arthur	36
Gold Hammer and Sickle Medal	27
Gold, Dr. Thomas	16
Goldstone, Calif	1, 6, 9, 29, 41, 45
Golovin Committee	61
Golovin, Dr. Nicholas E	43
Gordon, Lt. R. F. (USN)	22
Graham, Harold M	25
Gravity	63
Great Britain. (*See* United Kingdom.)	
Greater Los Angeles Press Club	68
Greb III	30, 83
Greenbelt, Maryland	3
Grissom, Capt. Virgil I. (USAF)	7, 33, 36, 58, 71
G. T. Schjeldahl Co	7, 21
Guam	40
Guidance	32, 34, 38, 42, 45, 47, 50, 53, 59, 70
Gulf of Mexico	17, 19
Gurtler-Hebert & Co	61
H-43-B (Huskie)	22
Hallman, Dr. Theodore M	26
Hagen, Dr. John P	50
Ham	4
Hardisty, Lt. Hunt (USN)	42
Harmon International Aviator's Trophy	27, 68
Havana, Cuba	34
Hawaii	27, 40, 46, 72, 75
Heartbeat	54

INDEX

	Page
Heat sink	32
Hector	7
Helicopter	13, 20, 22, 33, 48, 70
Helium	16, 54, 58
Helsinki, Finland	31
Henry Draper Medal	11
Hero of Socialist Labor	27
Herzfeld, Dr. Charles M	50
Hibbs, Dr. Albert R	65
High Altitude Observatory	73
Holloman Air Force Base, N. Mex	42
Holmes, D. Brainerd	48, 49, 59, 64, 77
Honolulu, Hawaii	40
Hooks, Maj. Gen. Daniel E	47
Hound Dog	61
Houston, Tex	47, 48, 73, 77
HSS-2	20, 22, 70, 78
Hughes Aircraft Co	3, 39, 44, 50
Humphrey, Senator Hubert H	15, 46
Hunguska River	67
Huntsville, Ala	9, 24, 27, 31, 39
Hurricane	45
Huskie. (See H-43-B.)	
Hustler. (See B-58.)	
Hyannis Port, Mass	37
Hydrogen	58, 62, 65
Hypersonic	28
Hypersonic Propulsion Research Laboratory	70
Ice runways	1
Ice Way, Project	1
Illinois, University of	33
Incentive Awards Committee (NASA)	26
India	68
Indian Ocean	10, 53
Indian River	39
Information retrieval	62, 76
Injun	30, 83
Institute of Flight Structures (Columbia University)	67
Institute of Industrial Science (Tokyo University)	44
Institute of Space Technology (Germany)	13
Institute of the Aerospace Sciences	23, 26, 74
Instituto Geofisico de Huancago (Peru)	51
Instrumentation Laboratory (MIT)	38
Intercontinental ballistic missile (ICBM)	65
International Academy of Astronautics	40, 52
International Aeronautical Federation (FAI)	23, 32, 34
International Astronautical Federation (IAF)	52
International Astronomical Union	40
International Committee on Geophysics	4, 53
International Conference on High Magnetic Fields	61
International Conference on Science and World Affairs	44
International Council of Scientific Unions (ICSU)	30, 56
International Exchange of Scientific Data	30
International Geophysical Year (IGY)	4, 53, 64
International Hypersonics Conference	40
International Meteorological Satellite Workshop	63
International programs	1, 2, 3, 6, 7, 26, 30, 31, 39, 63, 66
International Programs, Office of (NASA)	50, 52
International space agency	46
International space year	46
International Telecommunication Union	11
International Telephone & Telegraph Co. (ITT)	2
International Year of the Quiet Sun (IQSY)	4, 11, 30
Invention and Contribution Award (NASA)	64
"Investing in Scientific Progress"	32
Ionosphere	14, 17, 55
Ionosphere beacon satellite (S-45)	8, 17

INDEX

	Page
Ion propulsion	3, 32, 50
Iota I (*Spacecraft V*)	81
Iris	2
Irregular heartbeat	53
Israel	31, 63
Italian Air Force	71
Italy	2, 16
Ivans, W. S	8
Iven C. Kincheloe Memorial Award	50
Ivuna	65
Izvestia	20
J-2 (engine)	76
Jackass Flats, Nev	24, 62
James H. Wyld Memorial Medal	55
Japan	5, 8, 44
Ministry of Communications	1
Japanese Weather Bureau	35
Jastro, Dr. Robert	3, 74
Javelin	17, 26
Jet operations	75
Jet Propulsion Laboratory	1, 3, 9, 20, 25, 29, 32, 36, 41, 65, 66, 67
Jodrell Bank Experimental Station	21, 26, 40, 69, 76
Johannesburg, South Africa	45
Johns Hopkins University	25, 70
Johnson, Vice President Lyndon B	29, 32, 52, 54
Johnsville, Pa	14
Joint Chiefs of Staff	7
Joint Lunar Study Program Office (NASA-DOD)	28
Joint tenancy agreement (NASA-DOD)	32
Jucamarca Observatory (Peru)	51
Juno II	8, 17, 21, 82
Jupiter (IRBM)	16, 71
Kallmann-Biji, Dr. Hilde	60
Kamov	55
Kappa 1 (*Explorer X*)	81
Kappa-6	5
Kaman	22
Keldysh, Matislav	39, 48
Kennedy, President John F	1, 2, 4, 5, 8, 11, 12, 14, 15, 17, 20, 21, 22, 26, 27, 29, 32, 33, 34, 36, 38, 40, 41, 49, 58, 65, 68
Keplerian trajectories	69
Kerr, Senator Robert S	20, 28, 67
Khodarev, Jouli	26
Khrunichev, Lt. Gen. Mikhail V	15, 24
Khrushchev, Premier Nikita	6, 19, 27, 38, 44, 59
Kiwi B	24, 71
B-1A	62, 72
Komsomol Pravda	67
Kordylewski, K	28
Kozyrev, Nikolai	10
Krasnaya Zvezda. (*See* Red Star.)	
Kremlin	38
Kwajalein	64, 76
Lake Aral, U.S.S.R	23
Lambda 1 (*Discoverer XXIII*)	82
Laminar flow	24
Langley Aeronautical Laboratory	24, 75
Langley Air Force Base, Va	13
Langley Research Center	1, 3, 10, 18, 20, 21, 26, 30, 36, 37, 39, 41, 48, 61, 62, 64, 75
Langley, Samuel P	21
Large Launch Vehicle Planning Group	43
Launch Operations Directorate (NASA)	63
Lawrence Radiation Laboratories	9
Le Bourget, France	22, 24
Lederberg, Dr. Joshua	49
LeMay, Gen. Curtis E. (USAF)	21, 67
Lenin	35

	Page
Leventhal, Dr. Elliott C	49
Lewis Flight Propulsion Laboratory	34, 49
Lewis Research Center	5, 24, 30, 34, 49, 57, 59, 61, 65
L. G. Hanscom Field (Bedford, Mass.)	69
Liberty Bell 7 (MR-4)	33, 36, 84
Lick Observatory	25
Life Sciences Research Laboratory	2, 5
Lima, Peru	51
Lincoln Laboratory	16, 39, 62, 67
Lindbergh, Charles A	22
Liquid propellant	4, 5, 6, 31, 39, 49, 57, 61, 65, 76
"Liquid Thrust Vector Control"	75
Little Joe 5-B	17
Little Joe 6	10
Little Star	11
Liza	33
Lloyd, Capt. Bruce K. (USN)	70
Lockheed Aircraft Corp	3, 10, 37
Lockheed Missile and Space Co	9
Lofti	7, 14, 81
London Observatory	10
London, England	36, 53, 62
Long Island Sound	70
Los Alamos Scientific Laboratory	72
Los Angeles, Calif	22, 26, 60
Louis Bleriot Speed Trophy	24
Lovell, Sir Bernard	21, 26, 40
M-10 (U.S.S.R.)	47
McCauley, George D	59
McClure, Dr. Frank T	2
McClure, J. D	13
McCullough, Capt. James F. (USAF)	64
Mace-B	28
McDonnell Aircraft Corp	71
McKay, John B	5, 91, 92
McMeen, Capt. W. C. (USAF)	22
McNamara, Robert	14, 38, 59
Magnetic field	11, 12, 40, 61, 62, 72
Magnetogasdynamic electric rocket	13
Major Defense Systems Division (RCA)	48
Malinovsky, Marshal R. Y	57, 58
Management Council (NASA)	78
Manned Spacecraft Center (MSC)	43, 47, 48, 49, 71, 73, 77, 78
Manned space flight	3, 5, 19, 31, 33, 36, 37, 71, 74, 76
Manned Space Flight Programs, Office of (NASA)	48, 49, 61, 64, 77, 78
Mariner	32
R	78
Marquardt Corp	39, 70
Mars	4, 65
Marshall Space Flight Center (MSFC)	3, 4, 6, 9, 14, 15, 19, 24, 25, 28, 29, 31, 37, 40, 47, 50, 55, 57, 58, 62, 65, 72, 77, 78.
Martin-Marietta	5, 42, 76, 77
Maryland Academy of Sciences	56
Masevitch, Dr. Alla G	26
Mason-Rust	72
Massachusetts Institute of Technology (MIT)	1, 20, 29, 38, 39, 40, 61, 62
Materials Testing Laboratory (Hughes Aircraft Co.)	44
May Day	19
MB-3	43
Medal for Merit	34

	Page
Mercury (capsule)	4, 7, 10, 13, 17, 23, 26, 29, 33, 36, 46, 48, 57, 68, 70, 71
Mercury (planet)	40

Mercury-Atlas (MA):
- MA-2 ... 7
- MA-3 ... 17
- MA-4 ... 41, 46, 85
- MA-5 ... 63, 68, 71, 88
- MA-6 ... 71

Mercury, Project ... 1, 2, 7, 9, 19, 33, 40, 68, 71
    Tracking stations ... 13, 71
Mercury-Redstone (MR) ... 40
- MR-2 ... 4, 11, 58
- MR-3 ... 14, 17, 19, 25, 55, 82
- MR-4 ... 28, 29, 33, 48, 84

Mercury-Scout ... 59, 61
Meteorite ... 10, 13, 65, 67
Mice ... 14
Michigan, University of ... 11, 35, 40
Michoud Operations (NASA) ... 58, 61, 64, 72, 74
Michoud Ordnance Plant ... 45
Microbes ... 27
Micrometeorite ... 26, 29, 30, 41
Micrometeoroids ... 10
Micro-organisms ... 13
Midas III ... 32, 33, 83
Midas IV ... 56, 57, 58, 62, 87
Milford, Conn ... 22, 70
Miller, Representative George P ... 48
Millstone Hill ... 67
Minitrack station ... 3, 42
Minneapolis-Honeywell Regulator Co ... 50
Minnesota, University of ... 16, 45
Minuteman ... 5, 10, 21, 42, 52, 64, 75
Mississippi ... 58
Moffett Field, Calif ... 20
Molecule ... 44
Monkey ... 42, 44, 63, 68, 71, 76
Montgomery, Ala ... 47
Moon ... 16, 28, 46, 77
    Base on ... 73
    Exploration of ... 13, 38
    Manned landing on ... 4, 15, 22, 24, 26, 28, 29, 33, 41, 42, 68
    Orbit of ... 65, 71
    Soft landing on ... 3
    Surface of ... 1, 36
    Unmanned landing on ... 41, 78
Moscow, U.S.S.R ... 12, 15, 19, 24, 28, 30, 31, 33, 37, 38, 39, 53, 56, 58, 65
Moskalenko, Marshal Kiril S ... 46
Mu 1 (*Vostok I*) ... 82
Munich ... 78
Murphy, Maj. Elmer E. (USAF) ... 24
Murray, Ky ... 13
Nagoya University, Japan ... 44
Nagy, Bartholomew ... 65
Nancy, Hurricane ... 63
National Academy of Arts and Sciences ... 45
National Academy of Sciences ... 2, 10, 11, 16, 23, 25, 30, 38, 39, 48, 60, 61
National Advisory Committee on Aeronautics (NACA) ... 61, 75
National Aeronautic Association ... 20, 58, 66
National Aeronautics and Space Administration (NASA) ... 4, 5, 7, 8, 9, 11, 12, 13, 14, 16, 18, 19, 20, 24, 25, 26, 28, 29, 32, 33, 35, 36, 37, 40, 41, 42, 43, 44, 45, 46, 47, 48, 49, 50, 51, 52, 54, 55, 56, 57, 58, 59, 60, 61, 62, 63, 64, 65, 66, 68, 69, 71, 73, 74, 75, 78.

National Aeronautics and Space Administration (NASA)—Continued   Page
  Ames Research Center _____ 2, 5, 20, 39, 43
  Awards _____ 2, 20, 24, 26, 34, 59
  Contracts _____ 1, 3, 7, 9, 10, 13, 21, 24, 25, 27, 30, 31, 32, 35, 38, 39, 41, 42, 50, 57, 64, 68, 72, 76
  Deep Space Tracking Station _____ 1
  Goddard Institute for Space Studies _____ 3, 10, 21, 39, 74
  Goddard Space Flight Center _____ 3, 10, 12, 17, 25, 32, 39, 41, 58, 74
  International programs _____ 2, 3, 6, 7, 8, 10, 11, 13, 14, 16
  Jet Propulsion Laboratory _____ 1, 3, 9, 20, 25, 29, 32, 36, 41, 65, 66, 67
  Langley Research Center _____ 1, 3, 10, 18, 20, 21, 26, 30, 36, 37, 39, 41, 48, 61, 62, 64, 75
  Launch Operations Directorate _____ 63
  Launches:
    Probes _____ 56, 65
    Satellites _____ 7, 8, 11, 17, 21, 30, 32, 39, 65
    Sounding rockets _____ 1, 2, 5, 6, 11, 16, 26, 32, 39, 54, 56, 61, 74
    Spacecraft:
      Manned _____ 19, 33
      Unmanned _____ 4, 7, 46, 58, 68
  Lewis Research Center _____ 5, 24, 30, 34, 49, 59, 61, 65
  Manned Spacecraft Center _____ 43, 47, 48, 49, 71, 73, 77, 78
  Marshall Space Flight Center _____ 3, 4, 6, 9, 14, 15, 19, 24, 25, 28, 29, 31, 37, 40, 47, 50, 55, 57, 58, 62, 65, 72, 77, 78
  Space Nuclear Propulsion Office (SNPO) _____ 5, 9, 57, 71
  Space Task Group _____ 1, 7, 10, 20, 33, 37, 46, 48, 61
  Wallops Station _____ 2, 5, 6, 7, 10, 11, 16, 19, 20, 26, 30, 32, 39, 46, 54, 55, 56, 58, 70, 74
National Aeronautics and Space Council _____ 11, 14, 16, 17, 22, 27, 32, 39, 53
National Air Museum _____ 21, 57
National Aviation Facilities Experimental Center _____ 75
National Balloon Flights Facility _____ 16
National Bureau of Standards _____ 26, 50, 66, 69
National Center for Atmospheric Research (NCAR) _____ 16, 73
National Institutes of Health _____ 13, 25
National launch vehicle program _____ 7
National Meteorite Symposium _____ 10
National Meteorological Center _____ 45
National Nuclear Rocket Development Center _____ 42
National Operational Meteorological Satellite System _____ 51
National Press Club _____ 46
National Radio and Television Convention _____ 20
National Research Corp _____ 23, 33
National Rocket Club _____ 22
National Science Foundation _____ 11, 31, 32, 37, 45, 57, 66, 73
National Telemetering Conference _____ 21
National Weather Records Center _____ 31
Nature (periodical) _____ 65
Nausea _____ 53
Naval Missile Center _____ 8
Naval Research Laboratory (NRL) _____ 14, 22, 56
Naval Weapons Laboratory _____ 5
Navigation _____ 36, 38
Navigation satellite _____ 7, 30
Nerva _____ 25, 72
Netherlands West Indies _____ 35
Nevada _____ 47, 69
Nevada Test Site _____ 72
New Delhi, India _____ 68
Newell, Dr. Homer E _____ 49
New Hampshire, University of _____ 32, 39
New Orleans, La _____ 45, 50, 58, 72, 73, 74
New York, N.Y _____ 3, 21, 22, 33, 49, 54, 59
New York Times _____ 53
New York University _____ 65
New Zealand _____ 55
Nike _____ 61
Nike-Asp _____ 16

	Page
Nike-Cajun	2, 6, 28, 32, 35, 58, 66
Nike-Hercules	76
Nike-Zeus	47, 53, 64, 66, 68, 70, 72, 73, 74, 76
Nitrogen	16
Noble Co	47
NORAD. (*See* North American Defense Command.)	
North America	25
North American Aviation, Inc	3, 5, 8, 10, 13, 39, 46, 55, 57, 62, 68, 75
North American Defense Command (NORAD)	5, 31, 41
North American Newspaper Alliance	73
Northhampton College of Advanced Technology (United Kingdom)	36
Northrop Aircraft Corp	10, 24, 41, 54
Northwestern University	62
Norway	66
Nose cone	52, 55, 58
Nova	4, 29, 31, 37, 41, 54, 58
NRL	14, 22, 56
Nu 1 (*Explorer XI*)	82
Nuclear engine for rocket vehicle application (Nerva)	5, 25
Nuclear propulsion	6, 20, 24, 25, 65, 75
Nuclear reactors	28
Nuclear Vehicles Project Office (MSFC)	28
Octave Chanute Award	26
Office of Administration (NASA)	25
Office of Aerospace Research (OAR, USAF)	14, 48, 62, 68
Office of Launch Vehicle Programs (NASA)	9, 39
Office of Life Sciences (NASA)	14, 40
Office of Manned Space Flight (NASA)	48, 49, 61
Office of Naval Research (ONR)	11, 35, 63
Office of Programs (NASA)	25
Okinawa	37
Oklahoma	66
Omega 1 (*Discoverer XXX*)	85
Omicron 1 (*Transit IV-A*)	83
Optical solar simulation system	66
Order of Lenin	27
Order of the Badge of Honor	27
Order of the Red Banner of Labor	27
Order of the Red Star	27
Orgueil, France	10, 65
Oscar, Project	74
Oscar I	46, 73, 89
Ostrander, Maj. Gen. Don R. (USAF)	39
O'Sullivan, William J., Jr	64
Outstanding Leadership Medal (NASA)	24, 26, 59
Owens Valley, Calif	63
Oxygen	10, 13, 16, 38
Ozone	10
P-14 (*Explorer X*)	11, 12, 16, 43, 57, 81
P-21	56, 86
Pacific Missile Range (PMR)	8, 14, 32, 42, 47, 56, 58, 63, 64, 70
Pacific Ocean	17, 33, 37, 45, 46, 53, 55, 58, 64, 72, 75, 76
Palaemon	24, 29
Pamela, Hurricane	45
Paraglider	3
Pararescue	75
Paris, France	4, 22, 23, 24, 34
Paris International Air Show	23
Parkes, New South Wales	43, 69
Parsons, John F	43
Patent	69
Pearl River, Miss	58
Pearson, Drew	71
Pershing (missile)	68

	Page
PERT (program evaluation and review technique)	52
Petersen, Comdr. Forrest S. (USN)	38, 50, 91, 92
Pfenninger, Werner	24
Phantom II. (*See* F4H.)	
Phi 1 (*Ranger I*)	84
Philadelphia, Pa	48, 49, 56
Phoenix, Ariz	78
Physics	73
Pi 1 (*Discoverer XXVI*)	83
Pilot	3, 25, 33
Pilot-engineer	74
Planets	13, 38, 46, 49
Planning Task Force (NAS)	30
Plug-nozzle engine	20
Plum Brook	26
Pluto, Project	20, 69
Point Arguello, Calif	13, 65, 66, 70, 72, 76
Point Mogu, Calif	57, 61, 64, 72, 74, 76
Poland	28, 63
Polaris	1, 7, 9, 10, 20, 26, 32, 39, 50, 57, 62
Porous wing	24
Prather, Lt. Comdr. Victor G. (MC-USN)	19
Pratt & Whitney	4, 30, 65, 76
Pravda	3, 11, 27, 41, 42
President's Missile Sites Labor Commission	36
President's Safety Awards	28
Princeton, N.J	11
Princeton University	11, 67
Program evaluation and review technique. (*See* PERT.)	
Project Horizon	45
Propulsion Medal	55
Proton	35, 48, 70, 74
Psi 1 (*Discoverer XXIX*)	85
Queijo, M. J	36
Quiet Sun Year. (*See* International Year of the Quiet Sun.)	
Raborn, Vice Adm. William F., Jr. (USN)	20, 26
Radiation	17, 27, 30, 32, 40, 53
Radiation belt	37, 45, 74
Radiocarbon	61
Radio Corp. of America	11, 21, 30, 32, 43, 48, 50
Radioisotope	30
Radio Refractive Index Data Center	69
Radio Research Laboratories (Japan)	1
Radiotelescope	69, 76
Ramjet	20
Randolph Air Force Base, Tex	10
Randt, Dr. Clark	14
Ranger	32, 42, 56, 57
Ranger I	3, 9, 23, 37, 41, 42, 45, 84
Ranger II	45, 58, 62, 88
Rassegna International Electronic and Nuclear Fair	26
Reactor-in-flight-test (RIFT). (*See* RIFT.)	
Rebound, Project	6, 7, 31
"Reconnaissance, Mapping, and Geodetic Programs" (draft DOD directive)	12, 14
Recovery	68, 76
Area	19
Forces	65
Operation	64, 75
Recruiting of scientists	25, 61
Red Star (Kraznaya Zvezda)	33, 46
Redstone	30, 47
Reentry	16, 36, 43, 54, 61, 65, 68, 73, 74, 75, 76
Regulus II	19
Reichelderfer, Dr. F. W	17, 23, 63

	Page
Reid, Dr. Henry J. A	26, 30
Relay, Project	3, 6, 7, 21, 57, 78
Rendezvous	4, 46, 54
Research Analysis Corp	25
"Research and Development in Aeronautics"	52
Research medal	55
Respiration	54
Reuters	26
Rho I (*Tiros III*)	83
Rift (reactor-in-flight-test)	13, 72
Riley, Donald R	36
RL-10	65
Roadman, Dr. Charles A. (Brigadier General, USAFIMC)	1
"Road to Outer Space"	27
Robert H. Goddard Memorial Medal	55
Roberts, Dr. W. O	16
Rocket belt	25
Robinson, Lt. Col. Robert B. (USMC)	66
Rockefeller Foundation	69
Rocketdyne	43, 57, 76
Rogallo, Francis M	3
Roi-Namur	64
Roksonde	70
Rome, Italy	26
Rosman, N.C	74
Ross, Comdr. Malcolm D. (USNR)	19
Rotordyne	63
Roulstone, Comdr. E. J. (USN)	70
Rover, Project	5, 22, 25, 69
Royal Observatory (United Kingdom)	37
RS-70	77
Rudolph, Dr. Arthur	77
Runway slush	75
Rushworth, Maj. Robert (USAF)	53, 91, 92
Ryan Aeronautical Corp	3
S-I (Saturn stage)	57, 58
S-IB (Advanced Saturn stage)	72
S-3 (*Explorer XII*)	39, 40, 43, 57, 74, 84
S-45 (ionosphere beacon satellite)	8
S-55A (*Explorer XIII*)	41, 42, 85
SAC. (*See* Strategic Air Command.)	
Sacramento, Calif	26, 41
Sacramento Peak, N. Mex	48
Saenger, Dr. Eugen	63
SAGE Center	47
Sagen, Dr. Carl	12
St. Louis, Mo	56
Samos II	4, 80
San Antonio, Tex	25, 50
San Bernardino, Calif	76
Sandusky, Ohio	26
Sardinia	2
Saskatchewan, Canada	35
Satellite Communications Corp	67
Satellite	74
Second-generation	74
Saturn (booster)	2, 3, 4, 5, 6, 9, 15, 18, 19, 20, 24, 25, 29, 31, 37, 39, 41, 43, 45, 46, 50, 57, 58, 64, 72, 74, 76.
Schensted, Capt. W. C. (USAF)	46, 55
Schirra, Lt. Comdr. Walter M., Jr. (USN)	68
Schmitt, Dr. Otto	62
School of Aerospace Medicine (USAF)	25, 38
Schriever, Gen. Bernard A. (USAF)	14, 42, 49, 54, 76
Schwarzschild, Dr. Martin	11
Science Advisory Committee	45

	Page
"Science Organization and the President's Office"	27
Science Resources Planning Office (NSF)	57
Scientific Advisory Committee on Project Mercury (President's)	9
Scout	7, 9, 10, 16, 30, 65, 80, 85, 86
Seaborg, Dr. Glenn T	77
Sea King. (*See* HSS-2.)	
Seamans, Dr. Robert C	15, 42, 59, 73, 78
Sedov, Dr. Leonid I	52
"Self-Supporting Space Vehicle"	64
Seratov, U.S.S.R	23
"Series S Bonds for Space"	22
Settle, Vice Adm. T. G. W. (USN, retired)	36
Sharp, Dr. Edward R	24, 34
Shavit II	31
Sheingold, Dr. Leonard S	17, 51
Shea, Dr. Joseph F	77
Shepard, Lt. Comdr. Alan B., Jr. (USN)	7, 19, 20, 23, 25, 33, 55, 57, 59, 71
Siberia	67
Siepert, Albert F	25, 64
Sigma 1 (*Midas III*)	83
Silo (missile)	19, 42, 64, 75
Silverstein, Abe	49, 59
Sisakian, N. Mex	12
Skin divers	27, 44
Skybolt	1
Skylark	47, 61, 65
Skyshield II	55
Slayton, Maj. Donald (USAF)	68
Slider, Lt. J. (USAF)	38
Small-rocket lift device	25
Smelovka, U.S.S.R	23
Smithsonian Astrophysical Observatory	6, 15, 58
Smithsonian Institution	57
Snap	30
Sodium cloud	31, 46
Solar eclipse	6
Solar flare	8, 16, 48, 58, 72
Solar heat energy	1
Solar Laboratory (OAR)	48
Solar pressure	10, 39
Solar surface	11
Solar system	2, 21, 36
Solar thermionic power system	78
Solar winds	16
Solid Earth problems	30
Solid propellant	2, 3, 5, 6, 22, 24, 31, 37, 41, 47, 72
Sonic booms	60, 69
Sounding rockets	2, 5, 9, 26, 29, 31, 32, 35, 39, 60, 74
South Africa	45, 47
South Pole	70
Soviet Academy of Sciences	12, 21, 39, 45, 48, 52, 53
Soviet Air Force Day	31
Soviet Communist Party	35
Space condensers	30
Spacecraft	3, 32, 34, 35, 36, 44, 64
Spacecraft II	3
Spacecraft IV (Sputnik IX)	9, 81
Spacecraft V (Sputnik X)	11, 81
Space detection and tracking system (Spadats)	6, 31
Space exploration	25, 38, 45, 56
Impact of	66
Scientific need for	6
"Space Flight Report to the Nation"	53, 54
Space law	49, 67, 70
Space, military use of	53, 54, 62, 67

	Page
Space Nuclear Propulsion Office (SNPO)	5, 9, 57, 71
Space patrol	12
Space, peaceful use of	49, 53, 54, 56, 67, 70, 73
Space pilots	20, 74
Space probes	70, 74
Space program (United States)	13, 28, 29, 33, 38, 41, 44, 59, 62, 63, 68, 73, 77
Space race	77
Space radiators	30
Space Research Institute	27
Space science	75
Space Science Board (NAS)	2, 6, 13, 16, 23, 38
Space Sciences Committee (United Kingdom)	10
Space Sciences Laboratory (GE)	13, 68
Space Sciences, Office of, (NASA)	49, 61
Space Science Survey Team (Japan)	8
Space simulator	32, 38
Space station	62
Space surveillance system (Spasur)	5
Space Task Group	1, 7, 10, 20, 33, 37, 46, 48, 61
Space Technology Laboratory (STL)	66, 77
Speedball	64
Spinup rockets	50
Spurlock, Capt. D. A. (USMC)	78
Sputnik	15
Sputnik I	53
Sputnik IV (*Korable I*)	5, 8, 80
Sputnik V	6, 80
Sputnik IX (*Spacecraft V*)	11
Sputnik X	15
Stalingrad, U.S.S.R	45
Stanford University	49
State Committee for Coordinating Research Work (U.S.S.R.)	15
State, Department of	20, 53
State of the Union message	2, 4
State University of Iowa	39, 55
Steam motors	13
Sterilization of space vehicles	44, 72
Stevenson, Adlai E	37, 70
Storms, Harrison A., Jr	55
Stowe, Vt	44, 45
Strasbourg, Germany	4
Strategic Air Command (SAC)	55, 72
Stratolab High	17, 19
Stratoscope, Project	11
Strelka	3, 12
Stuhlinger, Dr. Ernst	65
Stuttgart Jet Propulsion Institute	63
Stuttgart, Germany	13
Suitland, Md	45
Sukhomlin, I	36
Sullivan, Comdr. Patrick L. (USN)	20, 22, 78
Sulzberger, C. L	44
Sun	15, 16, 20, 53
Sunnyvale, Calif	9, 37, 72
Supersonic transport	12, 45, 75
Supreme Soviet (U.S.S.R.)	27, 38
Surveyor	3
Sustained Superior Performance Award (NASA)	59
Sweden	41
Sycamore Canyon, Calif	57
Sydney, Australia	69
Sylvania Electronic Systems	17
Symington, Senator Stuart	11
Syncom, Project	57

	Page
T-38 (Talon)	10, 41, 54
Tass	3, 9, 21, 28, 38, 55, 58, 59
Tau 1 (*Vostok II*)	81
Telescope	25
Telstar, Project	57, 78
Tempe, Ariz	10
Temperatures	27, 35, 46, 54, 58, 60, 68, 70
Tennessee River	24
Texas	66
Textron	69
Theta 1 (*Spacecraft IV*)	81
Theodore Roosevelt Distinguished Service Award	59
Thermal arc-jet rocket engine	13
Thermopile	21
Thor	55, 57
Thor-Able-Star	7, 30, 63, 81, 83, 87
Thor-Agena B. (*See also* Agena B.)	27, 47, 80, 82, 83, 85, 86, 87, 89
Thor-Delta	32, 35, 39, 81, 83, 84
Thule, Greenland	1
Timken, H. A	22
Tiros	19, 30, 34, 57, 63
Tiros I	17, 31
Tiros II	1, 3, 5, 7, 9, 10, 11, 17, 21, 29, 35, 37, 43, 50, 57, 66
Tiros III	32, 33, 34, 35, 38, 43, 45, 57
Titan I	7, 34, 42, 45, 53, 57, 58, 66, 74
Titan II	3, 42, 53, 55, 70, 76, 77
Titov, Maj. Gherman S. (U.S.S.R.)	37, 38, 39, 41, 42, 47, 48, 53, 56
Tokaty, Dr. Grigori A	36
Tokyo University, Japan	44, 63
Topchiev, Alexander	12
Topside-sounder	11
Tory II-A-1	20
Traac	63, 87
Trailblazer	16, 74
Transit III-B	7, 14, 81
Transit IV-A	30, 83
Transit IV-B	63, 87
Tritium	58
Truman, President Harry S	34
TSX satellites	35
Tu-114 (Bear)	30, 36
Tullahoma, Tenn	13
Tulsa, Okla	23
Tulsa (Okla.) Chamber of Commerce	23
Tushino (airport)	31
Twenty-second Congress (U.S.S.R.)	35
Typhoons	40
U-2	33, 37
Ultraviolet	61
Underwater rocket launching	39
United Aircraft Corp	65
United Arab Republic	63
United Kingdom (U.K.)	3, 4, 6, 10, 14, 16, 26, 32, 44, 60, 61, 62
United Nations	23, 46, 49, 56, 70, 72
Committee on the Peaceful Uses of Outer Space	67, 72, 75
United Nations Conference	50
United Technology Corp	3, 37, 72, 75
Universe	63
University Corporation for Atmospheric Research	73
University of Illinois Observatory	33
Upsilon 1 (*Explorer XII*)	84
Uranium City, Canada	35
United States	10, 12, 13, 14, 23, 24, 26, 32, 44, 45, 47, 49, 54, 55, 57, 64, 65, 66, 67, 72, 73
U.S. Aerospace Industries in Europe	69
USAF Worldwide Information Conference	48

INDEX

	Page
U.S. Air Force	3, 5, 7, 8, 9, 13, 14, 16, 17, 20, 29, 31, 33, 37, 40, 50, 51, 54, 55, 60, 63, 64, 65, 69, 73, 74, 75, 76, 77
Air Defense Command	6
Air Force Ballistic Systems Division	39, 66
Air Force Missile Test Center	41
Air Force Proving Ground Command	20
Air Force Space Systems Division	25
Air Force Systems Command	14, 31, 47, 50
Air Research and Development Command	6, 14
Air Training Command	10, 47
Arnold Engineering Development Center	4, 35
Cambridge Research Laboratory	1, 15, 25, 29, 58, 69
Chief of Staff	6, 14, 21
Launches:	
Balloons	15
Missiles	3, 5, 7, 19, 21, 22, 28, 31, 34, 38, 42, 45, 46, 47, 52, 53, 57, 61, 63, 64, 66, 75
Satellites	4, 7, 25, 27, 31, 37, 42, 47, 55, 56, 57, 62, 63, 66, 73, 76
Sounding rockets	16, 20, 25, 29, 72
Space probes	70, 88
Office of Aerospace Research	14, 48, 62, 68
Strategic Air Command	55, 72
U.S. Army	14, 23, 25, 73
Chemical Corps	23
Corps of Engineers	25, 48, 73
Launches:	
Missiles	30, 64, 68, 70, 76
Target rocket	64
United States-Australian Ultraviolet Survey of the Southern Skies	47
U.S. Congress	2, 4, 20, 22, 34, 38, 41, 46, 51, 59
House of Representatives	16
Appropriations Committee	17, 20
Committee on Science and Astronautics	7, 8, 11, 15, 19, 47, 48, 52, 53, 54
Senate	6, 16, 44, 46
Committee on Aeronautical and Space Sciences	5, 6, 28, 29, 44
Committee on Government Operations	15, 27
Committee on Small Business, Subcommittee on Monopoly	62, 63
U.S. Marines	13
U.S. Navy	6, 7, 9, 12, 14, 16, 18, 20, 22, 29, 30, 32, 36, 55, 63, 67, 70, 75, 78
Launches:	
Balloons	17, 19, 65
Missiles	9, 16, 32, 39, 62
Satellites	7, 30, 63
Sounding rockets	61
U.S. News & World Report	29, 67
U.S.S. *Abraham Lincoln*	39
U.S.S. *Antietam*	17, 19
U.S.S. *Blandy*	58
U.S.S. *Decatur*	46
U.S.S. *Ethan Allen*	57, 62
U.S.S. *Observation Island*	9
U.S.S.R.	4, 6, 10, 11, 12, 15, 17, 21, 23, 24, 26, 28, 29, 30, 31, 32, 33, 36, 40, 42, 44, 45, 47, 48, 53, 55, 57, 58, 59, 63, 65, 71, 72, 73, 76
Launches	5, 6, 9, 15, 37, 46, 53, 55, 80, 82, 84
U.S.S. *Robert E. Lee*	16
U.S.S. *Stormes*	68
U.S. Weather Bureau	17, 23, 31, 33, 38, 63, 66
Vacuum	23, 27, 32, 44
Van Allen, Dr. James A	55, 56
Vandenberg Air Force Base, Calif	7, 19, 21, 27, 31, 33, 47, 72, 73
Vanguard	43
Vanguard I	10
Vanguard III	26
Varvarov, N	62
Vela, Project	66
Venice, Fla	47

INDEX

	Page
Venus	4, 9, 10, 12
Probe (U.S.S.R.)	6, 7, 9, 21, 26, 80
Probe (United States)	50, 78
Radar tracking of	15, 20, 26, 29, 43
Vernalis, Calif	15
Veronique (French rocket)	7, 62
Vertiplane	55
Vice President (United States)	14, 16, 17, 22
Vidsel, Sweden	39
Virginia Polytechnic Institute (VPI)	37
Virginia Peninsula	61
Vitro Engineering Co	42
Vnukovo Airfield (U.S.S.R.)	36
Volcano Ranch Cosmic Ray Research Center	10
Von Braun, Dr. Wernher	29, 50, 55
Von Kármán, Dr. Theodore	30
Voris, Dr. Frank B. (captain, MC-USN)	40
Vostok I	15, 21, 27, 48, 82
Vostok II	37, 38, 39, 41, 48, 53, 56, 57, 84
Wadachi, Kiyoo	35
Walence, Dr. Charles G	44
Walker, Joseph A	13, 22, 26, 27, 46, 50, 55, 68, 91, 92
Wallops Station, Virginia	2, 5, 6, 7, 10, 11, 16, 19, 20, 26, 30, 32, 39, 46, 54, 55, 56, 58, 70, 74
Warner, Dr. Brian	10
Washington, D.C.	6, 30, 33, 52, 59, 63, 75
Waterloo, Iowa	45
WB-66D	24
Weather satellites	1, 3, 4, 5, 7, 9, 10, 11, 17, 21, 22, 32, 33, 37, 63, 67, 70
Webb, James E	4, 6, 7, 8, 19, 22, 25, 28, 29, 31, 33, 34, 44, 46, 47, 48, 49, 54, 56, 57, 58, 59, 63, 66, 74
Weeksville, N.C.	21
Weightlessness	12, 17, 19, 38, 41, 53, 74
Welsh, Dr. Edward C	11, 53
Westbrook, Conn	22, 70
West coast	68, 76
Western European Conference	62
Western Union Co	55
West Ford, Mass	67
West Ford, Project	35, 39, 40, 50, 53, 56, 58, 62
West German post office	78
West Germany	31, 60, 62
Westinghouse Electric Corp	13, 25
Wheeler Dam	24, 37
White, Gen. Thomas D. (USAF)	6
White House	20, 34, 44, 46, 47, 53, 59
White, Maj. Robert (USAF)	5, 9, 16, 27, 29, 50, 54, 62, 63, 68, 91, 92
White Sands Missile Range (WSMR)	25, 44, 68, 70, 76
Wichita, Kans	1, 9
Wiesner, Dr. Jerome B	1, 2, 39
Wilmot Castle Co	24
Wilson, Capt. Jack (USAF)	31
Wilson, John L	57
Wilson, Lt. Gen. Roscoe C. (USAF)	31
Winckler, Dr. John R	16
Wind	35, 46, 70
Window	58
Windsor Locks, Conn	20, 78
Wind tunnel	21, 28, 29, 43
Winkfield, England	3
Witherspoon, Lt. Beverly W	20, 22
Wolf, Mrs. Constance	66
Woomera, Australia	6, 45, 47, 62
"Workshop: Telemetry in Europe"	21
World Magnetic Survey (WMS)	30
World Meteorological Day	11
World Meteorological Organization	6, 11, 23

	Page
Wright Brothers' Lecture	74
Wright-Patterson Air Force Base, Ohio	21
Wyatt, D. D.	25
X-15	5, 9, 12, 16, 22, 29, 46, 50, 53, 54, 59, 62, 63, 64, 65, 68, 75, 91, 92
No. 1	5, 8, 38
No. 2	6, 50
No. 3	75
X-248	43
Xi 1 (*Discoverer XXV*)	82
XLR-99 engine (X-15)	5, 12, 29, 38, 64, 75
XLR-115	65
X-ray	30
Yazdovsky, V. J.	53
YFNB barge	29
York, Dr. Herbert F.	9
Yost, Charles W.	67
Young, Lt. (jg.) B. R. (RIO)	22
Young, Robert B.	55
Yugoslavia	5
Zeta 1 (*Discoverer XXI*)	81
Zolotoukhin, A. A.	23, 63

○

www.ingramcontent.com/pod-product-compliance
Lightning Source LLC
Chambersburg PA
CBHW081728170526
45167CB00009B/3741